Biophotonics

Science and Technology

Biophotonics

Science and Technology

Yin Yeh

University of California, Davis, USA

V. V. Krishnan

California State University, Fresno
& University of California, Davis, USA

World Scientific

NEW JERSEY · LONDON · SINGAPORE · BEIJING · SHANGHAI · HONG KONG · TAIPEI · CHENNAI · TOKYO

Published by

World Scientific Publishing Co. Pte. Ltd.

5 Toh Tuck Link, Singapore 596224

USA office: 27 Warren Street, Suite 401-402, Hackensack, NJ 07601

UK office: 57 Shelton Street, Covent Garden, London WC2H 9HE

Library of Congress Cataloging-in-Publication Data

Names: Yeh, Yin, 1938–   author. | Krishnan, V. V. (Viswanathan Venkata), 1963–   author.
Title: Biophotonics : science and technology / Yin Yeh, V. V. Krishnan.
Description: New Jersey : World Scientific, 2018.
Identifiers: LCCN 2018000452 | ISBN 9789813235687 (hardcover)
Subjects: | MESH: Optical Imaging--methods | Molecular Imaging--methods |
    Microscopy--methods | Optics and Photonics--methods
Classification: LCC R857.O6 | NLM WN 195 | DDC 616.07/54--dc23
LC record available at https://lccn.loc.gov/2018000452

British Library Cataloguing-in-Publication Data

A catalogue record for this book is available from the British Library.

For any available supplementary material, please visit
http://www.worldscientific.com/worldscibooks/10.1142/10868#t=suppl

Desk Editor: Yu Shan Tay

Preface

This lecture volume is the culmination of the growth of a course I had initiated at the inception of the NSF-funded Center for Biophotonics Science and Technology (CBST) in 2002. The realization that interdisciplinary research such as biophotonics requires a knowledgeable group of participants who, in their fundamental training, already have some associated knowledge across the disciplines, be they physics, chemistry, biology, or the many avenues of engineering, is the inspiration for writing this volume. This course became the core curriculum for the Designated Emphasis in Biophotonics program at the University of California, Davis. During this 10-year period, I also presented shorter lecture courses based on this curriculum at the Chinese University of Hong Kong, and several Taiwanese institutions including the National Yang-Ming University, Chang Gung University, the Instrument Technology Research Center (ITRC), and the Academia Sinica.

The end of the era of the NSF support for the Center for Biophotonics and my retirement did not stop progress in this vast expanding field, as evidenced by the rapid development of ultrahigh resolution optical microscopy, including nonlinear optical excitation and sensing, and the important aspect of capturing dynamics of molecular function. Hence this volume came into fruition.

The reader will immediately note that the layout of this volume is to encourage gaining maximum interdisciplinary knowledge through in-depth involvement. After the introductory Chapters (2 and 3), the level of mixing between the sciences, all for the goal of gaining new knowledge in the basic understanding of nature, is maximized. These are exemplified in Chapters 4 through 7. I was introduced to the field of medical applications using photonic means by Dennis Matthews and Jim Boggan, co-Directors of CBST, and Brian Wilson, Director of CPIMA. Their strong influence is the reason that we conclude this volume with a chapter on medical applications using photonics means. National Science Foundation Manager of the CBST program, Denise Caldwell, had admonished the CBST group to "leave lasting legacies" of their work. This volume has attempted to summarize some of the work that the dedicated personnel of CBST had inspired.

The number of individuals that influenced the flavor of this volume and encouraged the publication of this set of notes is enormous. First I would like to acknowledge the effect of the late Professor Peter Debye on how to think about problems of nature and to learn to use all available resources (or invent new ones) in order to accomplish the goals. My compatriots in the teaching of this course, each of who had taught me greatly their diverse expertise and their love of scientific research, deserve much credit. They are Rodney Balhorn, Thomas Huser, Atul Parikh, and co-author of this volume, V.V. Krishnan. Without any of them, this work would not have the breadth it contains. My late colleagues and mentors Ronald J. Baskin, Herman Z. Cummins, and Robert E. Feeney, as well as current colleagues and friends Jiasen Chen, Arthur Chiou, Chien Chou, Ben Chu, Stephen Kowalczykowski, Chinlon Lin, and Din Ping Tsai all served as a major influence in the evolution of this volume. Thanks also to the work of postdoctoral fellows and former graduate students during their time associated with my lab. I would like to specifically acknowledge Piero Bianco, Robert Brown, Samantha Fore, Rick Lieber, Fred Milanovich, Bert Pinsky, Jim Selser, Sui Shen, Jamie Vesenka, and Huawen Wu. The editorial group at World Scientific has been very generous with their time and effort in pushing this work into its final form; we thank each and everyone of them.

Lastly, I would like to acknowledge the influence that my family has on my career. My father, Professor Chai Yeh, pointed me toward this direction of scientific pursuit. He published a volume entitled *Applied Photonics* when he was 82 years old. My brother, Jen, was the one who opened my eyes to biology by presenting me with a volume of James Watson's *Molecular Biology of the Gene* in 1970. My mother was my inspiration for staying in touch with the world, whether through work, art, or investments. My mother-in-law, Dr. Yun-teh Tang, continued her encouraging challenge for some tangible application of physics to conquer diseases. She lived to be a centurion. In heaven, I am sure she is happy to see these new developments for better health supported by biophotonics. Finally, this work cannot have been completed without the constant support and encouragement from my beloved wife and best friend, Elizabeth (Betty) Tang Yeh. It is her sense of staying involved, and involvement in depth as well as breadth, that created a model of meaningful engagement in life for me to emulate.

Yin Yeh
Davis, CA 95616

During my tenure at Lawrence Livermore National Laboratory and as an Adjunct Faculty at the Department of Applied Science at the University of California Davis, I had the opportunity to interact with the graduate students and teach part of the Optical Methods in Biology graduate course offered by Professor Yin Yeh. As a physicist by training, I have always been fascinated by the applications of optical methods in biology. When I was invited by Professor Yeh to work on this assignment, I did not want to miss the opportunity to enhance my understanding of the topic. Therefore, I would like to thank Professor Yeh for this chance to work with him again.

I also thank my family, particularly my children Maya and Haran for letting me split my time between Davis and Fresno and forgiving me for missing many important events in their life.

Finally, nothing would have been possible without Renuka (Renu), my dear wife for her unconditional love, unquestionable faith, countless professional and personal sacrifices and constant support in all my endeavors. You are always my MVP!

V.V. Krishnan
Davis CA, 95616 / Fresno CA, 93740

Contents

Chapter 1

Introduction to Biophotonics

1.1 Definition of Biophotonics

Biophotonics is not a word in the standard Merriam-Webster Dictionary. Yet this topic is as old as when the first optical microscopic image of an organism was taken. The meaning of "biophotonics" is essentially the use of optical or photonic means to examine, to track, and perhaps to control a biological process, at various levels of significant biology: molecular, cellular, tissue, and organismal level. The 20th century had been labeled the "Century of Physics," with the advances of now well-accepted principles of quantum physics paramount. Our level of understanding of matter is well couched with the knowledge of quantum mechanical principles, at the subatomic, the atomic, and the molecular levels. Complex molecular structures can be explained to a great degree by applying these principles of quantum physics. Over the latter half of the 20th century, we have also seen the discovery and development of quantum electro-optical devices, the laser, and its compatriot, the light-emitting diode (LED); ramifications of the range of applications for these quantum light sources seem to be unbounded. In this discourse, we will focus on how this type of light source has made an impact on our understanding of the new biology, arguably the area of most rapid advances for the 21st century in which we live. Indeed many consider this century the "Century of Biology"! Thus, biophotonics is ripe to be the interface between the significant developments in physics and engineering of the 20th century and the anticipated new findings in biology and its applications of the 21st century: medicine and the environment.

Biophotonics that combines physics, chemistry, engineering, and biology is highly interdisciplinary. At the very minimum, it is a field that brings in the physical knowledge of optical devices, such as light sources, detectors, and cameras, as well as special instruments such as the microscope to study biology. It also encompasses the knowledge of molecular structures based on our understanding of the physics and chemistry of molecules, couched in quantum mechanics. Finally, it forces us to be ever so innovative in engineering new devices and to process the massive array of data that comes with digital microscopy taken at the speed of live processes of biological significance. The illustration in **Figure 1.1** points to the essential needs of all of these background expertise, and the need to truly interface these fields of study so that a new "whole" is uniquely useful for enhancing our knowledge of biology and medicine.

Indeed, the timely award of the 2014 Nobel Prize in Chemistry to three pioneering researchers, William Moerner, Eric Betzig, and Stefan Hell, in the focus area of biophotonics stands tall in helping to claim that this field has finally arrived. In each of these three cases, the idea is to be able to visualize the biological system under investigation at a hitherto unheard-of level of spatial precision and to do this while the molecules are in as native a state of function as possible. These techniques of super-resolution optical microscopy are now leading the way to our understanding of the details of molecular function within a viable cell.

Simultaneously, the Nobel Prize for Physics in 2014 was awarded to Isamu Akasaki, Hiroshi Amano, and Shuji Nakamura for their invention of the efficient blue LEDs. This discovery has led the way to new engineering approaches in designing light sources to excite fluorophores efficiently. Albert Einstein developed the theory of light emission and absorption based on the now famous "A-B" rate coefficients in 1915. Though this discovery did not net him a Nobel Prize, we celebrated 2015 as the International

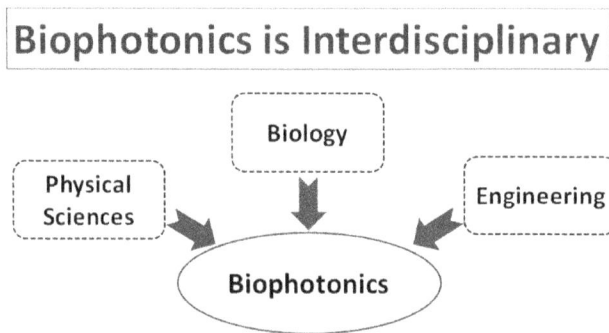

Figure 1.1.   An overview of interdisciplinary science of biophotonics.

Year of Light worldwide. The American Physical Society and the Biophysical Society had taken advantage of the year 2015 to proclaim the strong association of optical investigation with the significant, outstanding biophysical problems. The Congress on Laser and Electro-Optics (CLEO) had arranged a special symposium where not only these distinguished scientists participated, they were joined by Steve Chu, the 1997 Nobel Laureate in Physics for his work on optical molasses and the optical atomic trap, who is actively in the field of biophotonics.

In guiding this transition and spearheading the efforts to make unique contributions in the new field of biophotonics, the United States National Science Foundation (NSF) had funded the establishment of a Center for Biophotonics Science and Technology at the University of California, Davis, creating an interface between the Colleges of Engineering, Mathematics, Physical Sciences and Biological Sciences along with the Schools of Medicine and Veterinary Medicine. This center flourished from 2002–2012 and has now evolved into a new Center for Biophotonics at the University of California, Davis. Further driving this effort is the creation of the Biophotonics World website (https://www.biophotonics.world/), linking the scientific efforts of biophotonics research worldwide.

1.2   What are Photonic Processes?

Photonic processes are those interactions whereby electromagnetic disturbance, in the form of a wave or particulate, renders changes in the material's state and/or in the state of the incident electromagnetic radiation. We call these processes light-matter interactions. Because it is indeed the change in either the state of the matter or light that will inform us of the existence of such an interaction, we have strived to improve our approach to increase sensitivity for such detection, and in that process, gain new information about the matter being probed. So we consider the photonic methods as our "Biophotonics Tools Kit" (**Figure 1.2**).

Our toolkit includes, first of all, approaches to detect structures of matter (or molecules) by using just a single photon, capitalizing on the fact that electromagnetic radiation is a planar, transverse wave with specified polarization directions. These one-photon processes include absorption and birefringence; both linear and circularly polarized sources can provide unique information on the structure of matter being probed.

The Biophotonics Tools Kit

Absorption
Birefringence
CD/ORD

Optical
trapping

Structure

Dynamics

Function

Scattering
PCS
Raman

Fluorescence
FCS
FRET

Nonlinear
Optics
SHG/SFG
SRS/CARS

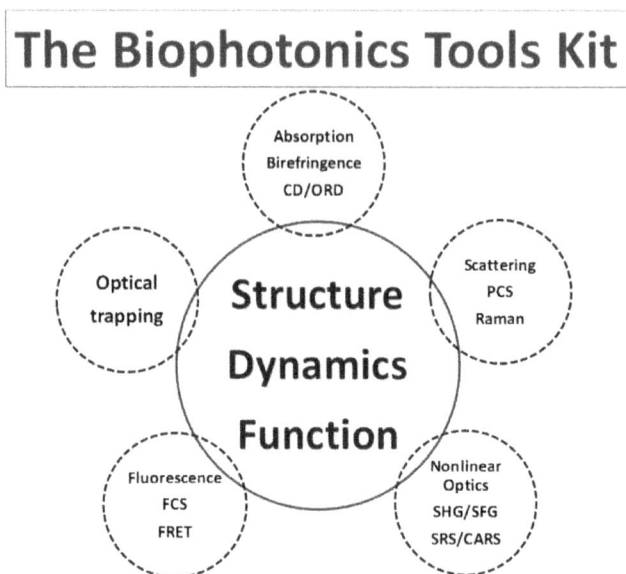

Figure 1.2. Biophotonics toolkit to understand structure and dynamics in biological processes.

Perhaps the most widely used light-matter interactions are those where two photons are involved. These may be of the phase-coherent processes such as light scattering, or phase-randomized, two-photon processes called fluorescence or phosphorescence. Unique identification of molecules can be achieved by designing fluorophores that have specific affinity for a specific molecule of interest. Indeed a major advance in fluorophore design was achieved by Roger Tsien, Osamu Shimomura, and Martin Chalfie who found that the green fluorescent protein (GFP) of the jellyfish (Aequorea Victoria) can be transfected into other organisms, thus rendering specific molecules of any organism fluorescently green. For this effort, they won the Nobel Prize in Chemistry in 2008.

Light scattering has been a standard-bearer for light-matter interaction studies since the time of Lord Rayleigh. Particle size differentials can easily be measured by light scattering methods, providing us with laymen arguments of why the sky is blue and a glass of milk is white when neither absorbs light. Peter Debye, who won the Nobel Prize in Chemistry in 1936 for his work on the scattering of short-wavelength light (X-rays) by molecules as a probe of molecular structure, continued to use the scattering of visible light for probing larger polymer structures. Going further into molecular spectroscopy, C.V. Raman discovered that molecular motion in the form of normal mechanical modes could create modulation signatures on the incident light shined onto the probed material, thus making these signatures the hallmark for the presence of specific molecules. For this discovery, he was awarded the Nobel Prize in Physics in 1930.

As is well-known, Charles Townes shared the 1964 Nobel Prize in Physics with Nicolay G. Basov and Aleksandr M. Prokhorov for fundamental work in quantum electronics, the fundamental precursor to the discovery of the laser. Equally important is the awarding of the Nobel Prize in Physics in 1981 to Arthur Schawlow for laser spectroscopy, shared by Nicolaas Bloembergen and Kai Siegbahn. In particular, Bloembergen started a new field of physics called nonlinear optics, whereby intense laser fields can elicit nonlinear materials responses, producing signals that are unique to the nature of the interaction and that of the material. This opened up the ability to capitalize these relatively weak material processes to identify material specificity more uniquely. We shall delve into nonlinear optics and related nonlinear optical microscopy in this volume.

Rounding out this toolbox, it is indeed the pioneering work of Steve Chu and colleague, Arthur Ashkin, that evolved the atomic trap into a tool for trapping larger molecules and for measuring forces at the molecular scale (pico-Newtons). This process depends on the radiation pressure gradient that can be exerted onto a well-controlled shaped particle of nearly micron size. Using this technique biologists are beginning to develop quantitative biological schemes of measurement, connecting force at this molecular scale to functions and structures.

With all these biophotonics tools in the toolbox of light, it is now time to move on to the next step and examine biological processes that can benefit from these tools.

1.3   Fundamental Biology Studies Need Photonics

In its most rudimentary view, biology describes an organism that has spatial extent, consumes matter for its chemical content, utilizes this chemical content by converting it into energy for the purpose of cellular metabolism, excretes waste matter, and is dynamic. We now know much more about the details of an organism, even one as simple as a single-celled bacterium or algae, or even a virus. Principally, the fundamental unit of the living entity is the cell. The cell, however, is not just a blob but is composed of millions of molecules each serving some special function to maintain life for the cell. As the technology in light microscopy improved, more and more details of a cell have become evident. Being true scientists, inquisitive individuals of the earlier era have been able to identify regions of the live cell and to assign differential functionality to them.

Cells all have a protective surface. Optically, using a transmission microscope of relatively low spatial resolution (e.g., 40×), one can show that the region of the cell cover is definitely distinct from that of the cell interior. Furthermore, in all eukaryotic cells, there are dense bodies within the interior of the cell. These are called nuclei. These cells are often dynamic in shape, frequently contorting themselves into highly convoluted shapes. The nuclear content of cells have become associated with genetic material, and the composition of the nucleus has been identified as deoxyribonucleic acid or DNA. By using another photonic method, X-ray diffraction, James D. Watson, Francis Crick and Maurice Wilkins (winners of the Nobel Prize in Physiology or Medicine in 1962) deciphered that the DNA within the cell consists of a double helix structure. Characterizing the DNA as a set of fundamental building blocks, composed of only four different DNA molecules, Adenine (A), Thymine (T), Guanine (G), and Cytosine (C), allowed scientists to postulate the origin of genetic transcription. The finding that in a double helix, the pairing of A–T and G–C is inviolable led to our understanding of how the double helix functions in the complementary template pairing scheme for the preservation of genetic material during transcription and cell division processes. Other optical methods, the optical rotatory dispersion and circular dichroism, allowed scientists to speculate how molecular melting can occur, leading to genetic mutations during cell division. Just how these activities take place in the cell and in the nucleus requires a higher spatial resolution optical microscope. Indeed, photons extending more into the X-ray wavelength domain play an important role in ascertaining that not only do proteins exist within the cell, but they also have major functions both on the cellular surface (membrane) as well as in the cell nucleus. Today we know many of these to be proteins. Linus Pauling's work on deciphering the molecular structures of proteins using X-ray diffraction methods won him the 1954 Nobel Prize in Chemistry. Now we know that there are many types of special proteins: membrane proteins, nuclear proteins, as well as proteins that self-assemble within the cytoplasm to form definitive molecular structures, foretelling the anticipated shapes and functions of the cells.

Under optical microscopic observation, with a resolution of approximately 1.0 micrometer (μm), cells of different shapes can be ascertained. The time trace of live cells presents another problem: Why do the cells change shape? They have been known to change shape in search of nutrients from external sources. The process of cytokinesisdescribes the movement of a cell. Voluntary and highly ordered rearrangement of proteins within a cell occurs in muscle cells upon a trigger signal, electrical or chemical. This suggests a sensory network within the cell for cell signaling and transduction. Molecular sources of such organized cellular activity in muscle cells have been identified as Ca^{++} ions. By using another optical method, fluorescence emission, it became possible to identify the movement of the Ca^{++} flux during nerve cell signaling processes. Although higher spatial resolution can be achieved by the use of electron microscopy (EM), the preparation protocol in EM usually precludes studying the cell domain in their functional state. Recent developments in cryo-EM have enhanced the value of this technique immensely. Appropriately, researchers of this technique development have been awarded the 2017 Nobel Award in Chemistry. Given the charge to the current surge of investigators that it would be most desirable to examine biological systems in their native functional state, it is the probing of the dynamics of the intracellular structure as they relate to functions of the cell that the field of biophotonics is tackling.

Where are the photonic needs? First of all, the cell membrane is composed of lipid molecules arranged into a molecular bilayer. This structure has a typical depth of approximately 5 nanometers (nm). The protein structures that dot the landscape of the membrane often are of dimensions of approximately 10 nm, and the membrane structure is now known to be highly heterogeneous. There is also the fact that the heterogeneity in the structure is highly dynamic, consistent with the dynamic movement (diffusion in two dimensions) on the membrane. The protein composition within the cytoplasm is completely dynamic; hardly is there a moment that a static structure can be ascertained. Even in the driven movements of collective muscle proteins, the head groups of the molecule within the myosin subfragment-I (S-1) that is the molecular motor is always in motion unless the cell is in the rigor state. Within the cell nucleus, nuclear proteins are constantly in search of DNA defects, in an effort to "search and destroy" before any defective genetic domain manifests as mutated genetic materi-

The Need for Biophotonics Tools

Membrane	Cytoplasm	Nucleus

• Structure
• Dynamics
• Pores
• Transporters
• Signaling
	• Intracellular
	  functions
	• Motility
	• Contractility
	• Metabolism
		• Information
		  Center
		• Replication
		• Transcription
		• Repair

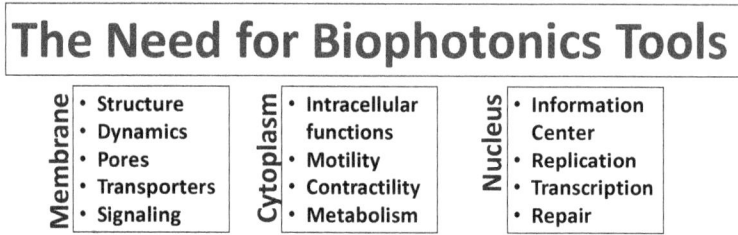

Figure 1.3. The role of biophotonics tools to understand the molecular cell.

als. How do we probe any and all of these functionalities of the cell? These are the outstanding types of questions that exist spanning the entire spectrum of biological molecules within a cell. As optical methods have evolved rapidly over the last 50 years, the application of new photonic methods to biology, hence biophotonics, becomes crucial for furthering the understanding of the dynamic assemblies of molecular networks within a cellular environment (**Figure 1.3**).

1.4   Applied Biology — Molecular Medicine Using Photonic Means

As humankind continues to inhabit the earth, the health of each continues to play a major role in the perpetuation of individuals and species. Healthcare in any society constitutes one of the largest components of that society's budget, and one is finding out that diseases are prevalent, knowing no geographic or ethnic boundaries. To name a few infectious diseases that the world had to encounter during the latter half of the 20th century, we recall the Ebola pandemic, the Acquired Immune Deficiency Syndrome (AIDS) virus, and the SARS epidemic are among these. As the healthcare protocols are getting more sophisticated worldwide, the life expectancies of humans are also getting longer, leading to a predominance of diseases related to aging. These include cardiovascular diseases such as heart attacks and strokes, metabolic diseases such as type-2 diabetes, Alzheimer's disease, and cancers affecting various organs.

Whereas in the past a physician would resort to either invasive procedures to deal with specific aspects of a disease, or find symptomatic relief to alleviate the suffering from these diseases, modern medicine has moved into the realm of molecular understanding of specific diseases. Routine blood tests now cover a battery of chemical panels that assess the quality of functions of organs at the molecular level. The aggressiveness of cancer cells is equally assessed at their molecular levels, leading to novel drugs that can alter the growth of cells to counter the assault of the invasive carcinogens. Stem cell research is progressing as possible modes of replacing very specific damaged domains of organs and whole blood. Abnormal cellular transport of small molecules and ions are confirmed to be the origins of many metabolic and nervous diseases, including type-2 diabetes and Alzheimer's disease. Medical researchers are searching for biomarkers specific to diseases, preferably at the smallest molecular level so as to catch diseases at their earliest stage and initiate possible interventions. The entire field of virology has evolved to characterize the different types of viral particles and perform dynamic tracking of how these particles specifically cause infections of cells. Applied biology and biomimetics have become the use of discovered biological principles to create molecular medicine for intervention against specific diseases.

The junction of biotechnology and photonics technology is biophotonics. This new field is developing a novel array of sampling methods to allow physicians to assess test results in real time as much as possible. Biomimetic approaches are being explored to control cellular dynamics and selective transport of small molecules into or out of cellular environments, often using the unique features of photonic approaches. Cancer therapy using photodynamic means are increasingly emphasizing nano-scale incursions via the use of highly specific antibodies with sensitive photonic responses. Site-directed

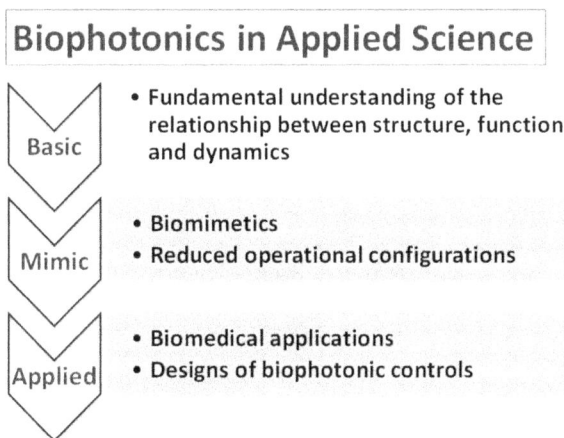

Figure 1.4.   Biophotonics from basic to applied sciences.

mutagenesis is allowing for possible solutions of eliminating genetic diseases, and incisions are being made at the genetic or DNA molecular level. The detailed visualization of all of these phenomena at molecular resolution, the time tracking of biochemical activities in the active intracellular environment, and the eventual control of molecular dynamics using photonic means constitute the rapidly evolving field of biophotonics (**Figure 1.4**).

1.5   Layout of This Volume

This lecture volume will present the topics consistent with the idea that we are discussing a highly interdisciplinary field. Thus, introductions to each of the necessary areas will be provided. The starting point will be the brief review of the field of optics (Chapter 2). This will be followed by a short introduction to the area of molecular physics, bringing about the essence of the quantum mechanical formulation of the molecular structure (Chapter 3). Within the same chapter, we shall see that it is indeed the unique molecular structure of certain molecules that allow for the development of specific molecules that are important to the molecules of life, as we know it. Thus, we will focus on the molecular structure of the simplest biological organism, the cell. This will bring in the intricacies of the types of molecules to protect the cell from insults, to perpetuate the generation of new cells of like kind, and to manage the operational necessity of such a complex environment.

Using these as core elements, Chapter 4 will delve into how currently a few representative, yet highly significant biological problems are being understood with the involvement of major optical tools. The represented problems are not exclusive in their application of optical tools, but they do show that the complexity of biology requires not one, but a range of tools for the solution of any of the mechanistic issues.

Chapters 5 and 6 form the major thrust in biophotonics, focusing on the current progress in using optical tools in biophysical research. These two chapters bring into play the most recent advances in optical microscopy, which has garnered several Nobel awards and certainly a few near-Nobel awards. Both labeled probing and label-free microscopy are discussed.

Chapter 7 discusses the expansion of optical microscopy to cover not only structural clarity of the cell but also how dynamics of molecules are being probed and how these motions are intricately driving the myriads of possible molecular structures to effect biological function.

The last chapter, Chapter 8, will allow us to examine how the basic science knowledge can be transferred to the realm of medical applications. In this chapter, we discuss how optical methodologies are being developed and evolved so as to not only be of diagnostic value but potentially significant therapeutic capability.

Chapter 2

Review of Electromagnetic Field Interaction with Matter

2.1 Electromagnetic Field, Intensity, and Photon Numbers

The classical description of light is that of an electromagnetic field that satisfies the physical laws of the Maxwell equations. Light waves are electromagnetic fields that oscillate transverse to the direction of propagation. Associated with this transverse wave is a wave frequency, ν, or wavelength, λ. These are related to each other by the speed of light, c. In a vacuum (Loudon, 1983; Pedrotti *et al.*, 2007),

$$c = \lambda\nu \qquad (2.1)$$

where c, the speed of light in vacuum, is a fundamental constant with a value of $c = 2.99 \times 10^8$ ms^{-1}. The propagation wavevector of light of frequency ν is denoted by $\mathbf{k}$, where its magnitude is given by

$$k = \frac{2\pi}{\lambda} = \frac{2\pi\nu}{c} = \frac{\omega}{c} . \qquad (2.2)$$

Thus a transverse electric field, $\mathbf{E}$, propagating along the z-direction with a wavevector $\mathbf{k}$ may be written as

$$E(z,t) = E_0 e^{-i(kz - \omega t)} . \qquad (2.3)$$

Maxwell equations allow two mutually perpendicular wave solutions, often called the **E**- and **H**-fields. Using vector notations for three-dimensional representation, $\mathbf{r} = [x, y, z]$ and $\mathbf{k} = [k_x, k_y, k_z]$, so that

$$\mathbf{E}(\mathbf{r}, t) = \mathbf{E}_0 e^{-i(\mathbf{k}\cdot\mathbf{r} - \omega t)}$$
$$\mathbf{H}(\mathbf{r}, t) = \mathbf{H}_0 e^{-i(\mathbf{k}\cdot\mathbf{r} - \omega t)} . \qquad (2.4)$$

Energy flux carried by this wave is referred to as the Poynting vector, **N**, given by

$$\mathbf{N} = \mathbf{E}^*\mathbf{H} = EH\mathbf{k}. \tag{2.5}$$

Here * represents the complex conjugate field. We see that energy is transmitted along the direction of the wavevector **k**.

Using another Maxwell equation to relate the **E** and **H** fields, one finds that, in vacuum,

$$\mathbf{N} = \varepsilon_0 |E|^2 \mathbf{k}. \tag{2.6}$$

So the energy carried by the field is proportional to the square modulus of the **E**-field ($|E|^2 = \mathbf{E}^*\mathbf{E}$). Thus, the Poynting vector has magnitude proportional to the intensity of the electric field! This observation from classical Maxwell equations has an important implication in the quantum field of radiation.

2.2 Classical or Quantum Mechanics Description of Interaction?

How does one decide if classical mechanics or quantum mechanics is needed for the description of matter or radiation or the interaction between light and matter? Quantum description of matter (including light) is couched in the formulations of *quantum mechanics*, where the classical Hamiltonian representing energy becomes the eigenvalue solution of the Schrödinger Equation (SE) (Atkins and Friedman, 2011). This SE is a wave equation; the solutions of which are wavefunctions. The eigensolutions are then associated with the discrete, quantum states consistent with the constraints of the SE, namely, force fields and boundary conditions. The observables are now described as the wavefunction densities associated with specific eigenenergies of that solution. The constituency of the matter is the "observable" defined by the expectation values of dynamic variables as related to the associated atoms and electrons. This fully quantum mechanical description of matter is, as it sounds, a complex problem. The field of *quantum chemistry* is devoted to the study of matter as these atoms come together to become molecules, and subsequently more complex materials of interest.

Our approach will be much less in-depth than that just described. We will bring in the necessary quantum mechanical description based on the parameters in question: *is quantum mechanics needed for that situation?*

When the wave equation is applied to matter, the result is that the measurable entity is represented by a wavepacket with a certain center wavelength and a frequency spread of this packet. More certainty of the particle's position means less spread in the wavepacket representation. These are the spatial Fourier components, each with slightly differing value in wavelength and wavevector. Since Heisenberg's *uncertainty principle* applies to all matter and light, a wavepacket that represents a definable matter state must subscribe to the uncertainty allowed by that principle. de Broglie had set forth to formulate the concept now called the *de Broglie wavelength* for matter.

The idea behind this concept is that any particle at above Kelvin zero temperature has some kinetic energy (KE): calling this KE just E

$$E = \frac{1}{2}mv^2.$$
(2.7)

The de Broglie hypothesis is that matter will have a characteristic wavelength depending on its energy. de Broglie used the equation:

$$\lambda_B = \frac{h}{p} = \frac{h}{mv}$$
(2.8)

where λ_B is the material wavelength, and is related to Planck's constant, h, divided by the linear momentum of the particle, p. Since classically, $p = mv$, with v being the velocity of the particle of mass m, immediately one sees that larger particles will lead to shorter λ_B.

Using the classical equi-partition theorem that says a particle will have $\frac{3}{2}k_BT$ (k_B is Boltzmann constant, and T is the temperature) of kinetic energy for the three degrees of freedom if moving unhindered in any direction, we can write

$$E = \frac{1}{2}mv^2 = \frac{3}{2}k_BT.$$
(2.9)

From which we can write:

$$mv = \sqrt{3mk_BT}.$$
(2.10)

Substituting this into the de Broglie equation (Eq. (2.8)), we then have

$$\lambda_B = \frac{h}{\sqrt{3mk_BT}}.$$
(2.11)

Now let us put in a few typical examples to see where these λ_B values are in comparison with the optical wavelengths of our interest.

(1) Free electron at room temperature

$$h = 6.626 \times 10^{-34} \text{ Js}; \quad k_B = 1.380 \times 10^{-23} \text{ JK}^{-1};$$
$$m_e = 9.109 \times 10^{-31} \text{ kg}; \quad T = 300°\text{K}$$
$$\lambda_B = (6.626 \times 10^{-34} \text{ Js})/\sqrt{3 \times 9.109 \times 10^{-31} \text{ kg} \times 1.380 \times 10^{-23} \text{ JK}^{-1} \times 300°\text{K}}$$
$$= 6.23 \times 10^{-9} \text{ m}$$
$$= 6.23 \text{ nm}.$$

(2) We now look at a larger mass particle, such as fullerene, C_{60}.

Here, the mass of the particle is that of 60 carbon atoms. This is $m_{C_{60}} = 60 \times 1.674 \times 10^{-27}$ kg $\approx 1.0 \times 10^{-25}$ kg. Holding all other quantities constant, we can write

$$\lambda_B = (6.626 \times 10^{-34} \text{ Js})/\sqrt{3 \times 1.0 \times 10^{-25} \text{ kg} \times 1.380 \times 10^{-23} \text{ JK}^{-1} \times 300°\text{K}}$$
$$= 1.9 \times 10^{-11} \text{ m}$$
$$= 19 \text{ pm}.$$

We next examine the test of the need to apply quantum principles in both cases. For the electron, since the de Broglie wavelength is in the nanometer regime, it is very comparable to the length parameters of atoms and molecules. This means that in the electron-atom interaction, consistent with *Young's principle of light diffraction*, electron waves of wavelength λ_B will be diffracted by the atoms. Thus, in the interaction problem of electron and atoms, quantum mechanics will have to be used. On the other hand, for the fullerene particle, we see that the de Broglie wavelength is much smaller than the atomic dimensions. Thus, diffraction of fullerenes by atoms is negligible, and we can consider this interaction to be a classical, ballistic, Newtonian mechanics problem. As a point of information, heated fullerenes have been observed to diffract in the predominantly forward direction from grating structures with spacing at 100 nm (Arndt *et al.*, 1999). As a general rule though, larger molecular entities can be treated in the classical correspondence limit of quantum mechanics.

2.3   The Radiation Field and Its Quantum Description

An interesting, totally solvable problem in quantum mechanics is that of a harmonic oscillator. We can write the classical Hamiltonian, H for the oscillator as

$$\hat{H} = KE + PE = \frac{\hat{p}^2}{2m} + \frac{k\hat{x}^2}{2} \tag{2.12}$$

where $\hat{p}$ and $\hat{x}$ are the quantum mechanical *operators* for dynamic variables momentum and position respectively. Using

$$\hat{p} = -i\hbar\frac{d}{dx} \quad \text{and} \quad \hat{x} = x \tag{2.13}$$

where $\hbar = \frac{h}{2\pi}$. In the Schrödinger's picture, the *time-independent* operator equation becomes

$$-\frac{\hbar^2}{2m}\frac{d^2}{dx^2}\psi_n(x) + \frac{1}{2}kx^2\psi_n(x) = E_n\psi_n(x). \tag{2.14}$$

This wave equation for $\psi_n(x)$ has well-defined energy solutions E_n called the Eigen (or characteristic) energies of the system.

$$E_n = \left(n + \frac{1}{2}\right)h\nu_n. \tag{2.15}$$

Moreover, the corresponding Eigen wavefunctions are $\psi_n(x)$. The evenly-spaced energies levels in the quantum mechanical description of the harmonic oscillator constitute a unique feature of this system. The fundamental oscillation frequency of the mode is ν_0. As n increases, $E_{n+1} - E_n = h\nu_0$ remains a constant of the system with this mode frequency. The nth eigenstate of an oscillator with mode ν_0 may be described as one with n of these $h\nu_0$ fundamental units of energy. Also, for higher n states, the oscillatory wave solutions become more and more like the classical solution, exhibiting dwell points (highest probability density) $|\psi_n(x)|^2$ at the end points of x-traversal and lowest probability density at $x = 0$. When compared to classical mechanics, hence $n \to \infty$, the solution of the classical harmonic oscillator problem is indeed also an oscillatory motion about the equilibrium point $x = 0$.

We next consider our electromagnetic wave solution. From Maxwell's equations, we can develop an equation to describe the classical electromagnetic wave $E(x,t)$ in space and time. Focusing once again on the time-dependent solution of that wave equation, we have the one-dimensional wave equation written as

$$\left(\frac{d^2}{dx^2} - \frac{1}{c^2}\frac{d^2}{dt^2}\right)E(x,t) = 0. \tag{2.16}$$

Using the solution of this classical wave equation (Eq. (2.3)), $E(x,t) = E_0 e^{-i(kx-\omega t)}$, where $k = 2\pi/\lambda$ and $\omega = ck$ with c as the speed of light in vacuum. The time-independent equation (Eq. (2.16)) then becomes

$$\left(\frac{d^2}{dx^2} - k^2\right) E_n(x) = 0 \tag{2.17}$$

where $k = \omega/c$.

These equations show that the solutions of the classical wave equation in time-independent mode are identical to the time-independent solutions of the Schrödinger wave equation for the harmonic oscillator! If we relate the trajectory equation of motion of the mechanical oscillator as the electromagnetic field equations, outside some scaling factors, the description of the oscillator is identical to that of the EM field. If one attributes a fundamental entity $h\nu$ to be the ground state energy of an oscillator (either mechanical or EM field) with frequency ν_0, we can then call this a *photon* of energy quantum unit $h\nu$. So, from the expression above, E_n can be n fundamental units of photons, each of energy $h\nu$ without any loss of generality. In this way, we can relate the classical electric field intensity to the energy of a well-defined number of photons. In our analysis of light-matter interaction, there will be occasions for us to use the classical approach, and there are other situations where the quantum description of light, namely photons, is more useful.

2.4 One-photon Matter Interaction

2.4.1 *Linear absorption*

Assuming that the particle of interest is indeed with a de Broglie wavelength that is sufficiently small so that under normal optical radiation interaction, it can be considered a continuum, we can then describe the interaction of light with such a particle by a semi-classical theory. Let us first consider this particle to be an atom. The semi-classical idea of an atom is that there is a heavy nucleus surrounded by electrons. Without an electric field on this atom, the atom is in a stable and charge-neutral state. Upon applying the electric field of value $\mathbf{E}$ on this particle, both the positively charged nucleus and negatively charged electron(s) are subjected to motion due to the field by $\mathbf{F} = -e\mathbf{E}$. Because the nucleus is at a minimum approximately 2,000 times heavier than the electron, the movement of this nucleus is much less than that of the electron. Thus there is a relative displacement between these two particles. In this classical picture, the displacement of charged particles relative to each other sets up a dipole field, here totally induced by the presence of the incident electric field, **E**. We can describe the relative displacement by a mechanical oscillator equation, here in just one dimension. That is

$$m\frac{d^2x}{dt^2} + m\gamma\frac{dx}{dt} + m\omega_0^2 x = -eE \tag{2.18}$$

with

$$E(t) = E_0 e^{-i\omega t} . \tag{2.19}$$

Here, x is the relative displacement of the (valence) electron from the nucleus, ω_0 is the characteristic frequency of the mechanical oscillator that we have used to represent the particle of interest. The electron is assumed to have typical charge, e, and mass, m. We insert a classical phenomenological dissipation factor, γ, (proportional to the velocity of motion) to represent all aspects of mechanical damping resistance. The applied field has the field strength, E.

The solution to this mechanical harmonic oscillator equation follows as

$$x(t) = x_0 e^{-i\omega t} \tag{2.20}$$

$$x_0 = \frac{-(eE_0/m)}{(\omega_0^2 - \omega^2) - i\omega\gamma} \tag{2.21}$$

$$p = -ex = -ex_0 e^{-i\omega t} = \frac{(e^2/m)[(\omega_0^2 - \omega^2) + i\omega\gamma]}{(\omega_0^2 - \omega^2)^2 + (\omega\gamma)^2} E_0 e^{-i\omega t} \tag{2.22}$$

$$p = \alpha E . \tag{2.23}$$

So that the induced dipole field, p, exhibits a complex response function.

Now we assume in this classical picture, each of the matter systems is identical to the one we just described, and all of these oscillators are non-interacting with each other, then the response of the total system composed of N particles in a volume space of V becomes

$$P = \left(\frac{N}{V}\right)p = \varepsilon_0 \chi E_0 \tag{2.24}$$

$$\chi = \frac{(Ne^2/\varepsilon_0 mV)[(\omega_0^2 - \omega^2) + i\omega\gamma]}{(\omega_0^2 - \omega^2) + (\omega\gamma)^2} = \chi' + i\chi'' \tag{2.25}$$

$$\chi' = \frac{(Ne^2/\varepsilon_0 mV)(\omega_0^2 - \omega^2)}{(\omega_0^2 - \omega^2) + (\omega\gamma)^2} \approx \frac{(Ne^2/\varepsilon_0 mV)(\omega_0 - \omega)/2\omega_0}{(\omega_0^2 - \omega^2) + (\gamma/2)^2} \qquad (2.26)$$

$$\chi'' = \frac{(Ne^2/\varepsilon_0 mV)\omega\gamma}{(\omega_0^2 - \omega^2) + (\omega\gamma)^2} \approx \frac{(Ne^2/\varepsilon_0 mV)\gamma/4\omega_0}{(\omega_0^2 - \omega^2) + (\gamma/2)^2} . \qquad (2.27)$$

In the above relationships, we have called the macroscopic field of induced dipoles the polarization field, P. The macroscopic susceptibility is now called χ, where the real and imaginary components of the susceptibility are given by χ' and χ''. From these two equations, we can see that the macroscopic response function has a resonant denominator in the form of $(\omega_0^2 - \omega^2) + (\gamma/2)^2$. Thus, if the applied E field has a frequency of $\omega \sim \omega_0$, then the denominator is nearly zero (besides the damping factor). The peaking of χ'' is characteristic of such a resonant behavior.

Note also the resonant feature of χ' is totally different. In this case, χ' becomes 0 when resonance is reached (**Figure 2.1**).

How we relate these complex expressions to measurable quantities requires that we consider whether the material essentially a dielectric. From classical Maxwell equations, we recall that the displacement field, **D**, is related to the applied field, **E**, by: $\mathbf{E} \Rightarrow \mathbf{D} = \mathbf{E} + \mathbf{P} = (1 + \varepsilon_0\chi)\mathbf{E} = \varepsilon\mathbf{E}$. Here, ε_0 is the electric permeability in a vacuum. Given our description of the polarization field, **P**, above, we can substitute and write

$$\varepsilon = (1 + \varepsilon_0(\chi' + i\chi'')) . \qquad (2.28)$$

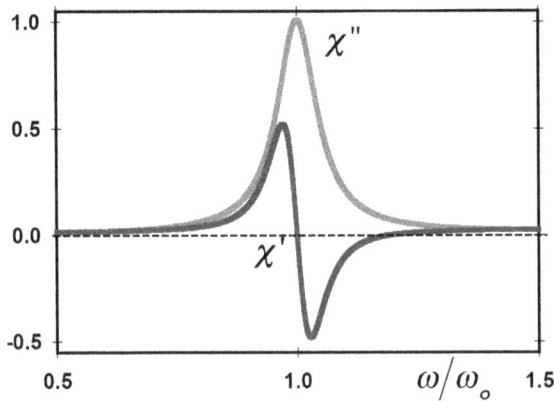

Figure 2.1.   Normalized absorptive (χ'') and dispersive (χ') components as a function of ω/ω_0.

In a non-magnetic dielectric, the index of refraction of the medium can be related to the complex dielectric permeability: $\varepsilon = n^2 = (\eta + i\kappa)^2$. We now have, for the wavevector $\mathbf{k}$, in the dielectric, given by $|\mathbf{k}| = nk_0 = (\eta + i\kappa)k_0$ where k_0 is the magnitude of the wavevector in a vacuum. We next note that the wavevector, $\mathbf{k}$, in the dielectric is also a complex quantity, and can be easily related to the susceptibility relationship from before. Specifically, the propagation wavevector in the dielectric is given by ηk_0, and the field's attention length, r_d, is related to

$$r_d^{-1} = 2\kappa k_0 \,. \tag{2.29}$$

Doing a little math, we arrive at

$$\left(\frac{k}{k_0}\right)^2 = (\eta + i\kappa)^2 = \eta^2 - \kappa^2 + 2i\eta\kappa = 1 + \chi' + i\chi'' \tag{2.30}$$

$$\therefore \; 1 + \chi' = \eta^2 - \kappa^2 \quad \text{and} \quad \chi'' = 2\eta\kappa \,. \tag{2.31}$$

Since both χ' and χ'' are quantities with unique frequency dependence, we expect η and κ to exhibit correspondingly similar dependence. Here, we note the dispersive and resonance features of χ' and χ'' are reflected in the complex index of refraction as well. In particular, the $\kappa(\omega)$ relationship has a maximum at $\omega/\omega_0 = 1.0$, representing maximum absorption of the incident light should the wavelength of the incident light be exactly the same as the resonance frequency. This is the optical resonance condition. Thus a measure of the spectroscopic absorption peak is providing information about the natural resonance frequency of the material under investigation. The field of optical absorption spectroscopy is based on this single concept.

We also note that the real part of the index of refraction, η, has a unique dip in its profile at the point of resonance, $(\omega/\omega_0) = 1.0$. For values of $(\omega/\omega_0) < 1.0$, η is always above 1.0, the value of the index of refraction in a vacuum. For values of $(\omega/\omega_0) > 1.0$, the η value is always < 1.0. These are called the normal dispersion ranges for light propagating in a dielectric medium (**Figure 2.2**). The transition region is as broad as the frequency width of the resonance process. Within this transition region, the dispersion curve has a negative slope, which is considered anomalous. Thus encountering an anomalous dispersion region is an indication that there is resonance. Using either dispersion or absorption as measuring tools, changes in those quantities are sensitive measures of the presence of a specific substance, and changes in the absorption or dispersion signals indicate that the material is undergoing change. We shall go into these as we encounter the biomaterials.

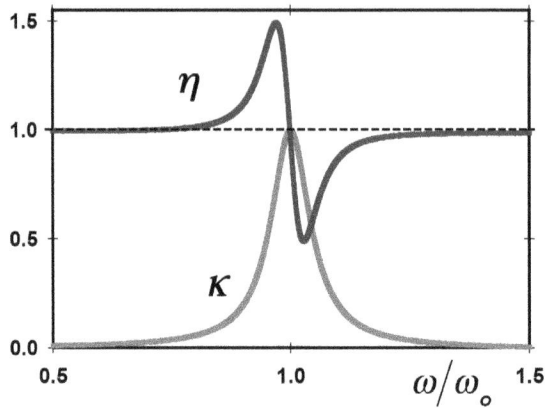

Figure 2.2.   Plot of the absorptive (κ) and dispersive (η) components as a function of ω/ω_0. Horizontal line on 1.0 indicates the vacuum index.

2.4.2 *Birefringence and dichroism*

The above description of the macroscopic dielectric medium is developed mainly for an isotropic medium. That is, the medium's response to the applied electric field is always and completely in the direction of the polarization of the incident field. However, even for totally isotropic dielectric medium, the response of the medium at boundaries depends on the nature of the applied field polarization. In elementary courses of electricity and magnetism, we learned about the two mutually perpendicular polarizations of the electromagnetic field, transverse electric (TE) field, and transverse magnetic (TM) field. For even an isotropic dielectric medium, the incident electric field at an angle θ will generate different responses in the reflected light and transmitted light, depending on the field being TE or TM polarization. We will exploit this characteristic in a later section.

For a medium of dielectric nature, the composition may be birefringent or dichroic, indicating that the response of the material along one polarization direction is different from that of the other. A simple example of a birefringent material is a crystal that has more than one optical axis. Consider a uniaxial crystal. This material in its crystalline form has one symmetry axis in its material makeup in such a way that if light were to be propagating along that axis, called the optical axis, the two transverse polarizations, TE and TM, experience the same index of refraction η. However, since a crystal is a three-dimensional object, propagating the incident

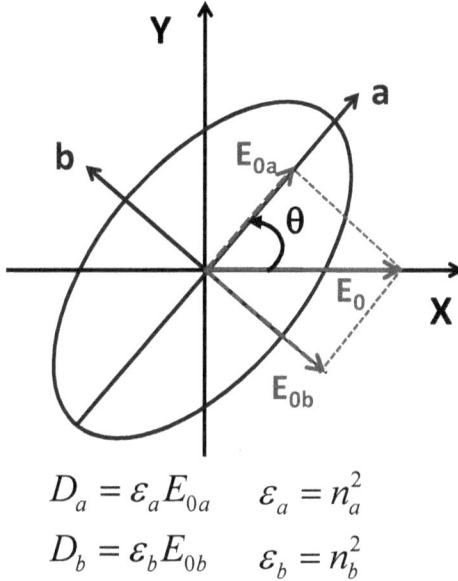

$$D_a = \varepsilon_a E_{0a} \qquad \varepsilon_a = n_a^2$$
$$D_b = \varepsilon_b E_{0b} \qquad \varepsilon_b = n_b^2$$

Figure 2.3. Propagation of light E_0 along x-axis in a dielectric medium. Uni-axial material oriented at angle θ with respect to x-axis. The indicial ellipsoid has electric permeabilities ε_a and ε_b.

beam of light along any other axis that is not the optical axis is certainly possible. In that case, assume that the optical axis is tilted from the laboratory axis (x) by an angle of θ (**Figure 2.3**) and we impose a field normal to the plane of the paper with its E_0 directed along x. The projection of E_0 along the body axes, a and b, would be $E_a = E_0 \cos\theta$ and $E_b = E_0 \sin\theta$. The response oscillations along those directions would be $D_a = \varepsilon_a E_a$ and $D_b = \varepsilon_b E_b$. Thus, the mutually perpendicular response fields experience different indices of refraction. As we had noted that the dielectric susceptibility is complex in an absorbing medium, the response functions D_a and D_b will show phase shift as well as amplitude differences. The result is that the observed field, D, now contains the information about the material, showing a degree of optical anisotropy or a change in the optical polarization orientation. The degree of optical anisotropy is given by the shape of the indicial ellipsoid, or in the example, $\tan\varphi \equiv \sqrt{\frac{\varepsilon_b}{\varepsilon_a}}$. In Maxwell equation language, the equation of $\mathbf{D} = \varepsilon\mathbf{E}$ now becomes a tensor expression

$$\boldsymbol{D} = \boldsymbol{\varepsilon}\cdot\boldsymbol{E} \qquad (2.32)$$

where the ε is a 3×3 dielectric permeability tensor quantity. In its general form, where ε has nine non-vanishing components, the displacement field **D** will be characteristically in a different orientation from the applied field **E**. If one transforms to the principal axes of the crystal, the off-diagonal contributions become zero, and we have three non-zero elements for the tensor, ε.

$$\varepsilon = \begin{pmatrix} \varepsilon_{11} & 0 & 0 \\ 0 & \varepsilon_{22} & 0 \\ 0 & 0 & \varepsilon_{33} \end{pmatrix}. \tag{2.33}$$

These are the $\varepsilon_{11}, \varepsilon_{22}, \varepsilon_{33}$ elements, corresponding to the principal components of the crystal tensor along the symmetry directions $\{x_1, x_2, x_3\}$. For uniaxial crystals, we can impose the further conditions that $\varepsilon_{11} = \varepsilon_{22} = \varepsilon_b$ and $\varepsilon_{33} = \varepsilon_a$. Now, if we relate the index of refraction to the dielectric permeability elements, we can write $\eta^2 = \varepsilon$, for these non-magnetic materials. We see that there is a difference in the real part of the index of refraction in these crystals, hence light propagating along these directions will travel at different velocities $(v = c/\eta)$. Such a birefringent crystal will create a phase lag along one of the polarization directions, and the resultant field is now elliptically polarized even though we had sent a single, linearly polarized field into the material. Thus a measurement of the degree of birefringent field provides information about the material that had generated the birefringence.

Because the fields that have interacted with the material will have different phases in their propagation constants, these methods become sensitive monitors of the presence of such materials. So even in the simplest experiment of a transmission study, where light passes through a medium of unknown matter, aspects of the unknown dielectric matter will be revealed. The techniques of optical absorption near a resonance of the material are usually used to study the content of such materials. Even for systems far removed from the resonant conditions, it is possible to gain insights from the differential phase lags.

Experimentally, one would use as an incident light source either linearly polarized or circularly polarized light. After passing through the medium of interest, one measures the difference in the phase by the change in optical ellipticity or a rotation of the plane of polarization. These differential measurements are very sensitive to probing small changes in minute concentrations of solution matters. *Linear birefringence* and *linear dichroism* then become tools for probing materials with these features.

2.4.3 *Circular dichroism and rotatory dispersion*

It is possible that materials of larger spatial dimension do not necessarily assemble to become crystals but nonetheless exhibit certain levels of ordered asymmetry. Such may be the case of polymeric molecules linked together as in DNA or proteins, as we shall discuss in the next chapter. It is possible that the local order of these polymeric arrangements can lead to optical anisotropy. Consider the case of electric dipoles linearly ordered to form a helix. Such an ordered system will exhibit rotational asymmetry in the form of circular birefringence (often called optical rotatory dispersion, ORD) and circular dichroism (CD). The degree of optical rotation that can be measured is then an indication of the material's helical nature, with the sense of the helix, left or right, clearly discernable. Very often in the probing of large helical, polymeric molecules, the use of circularly polarized light to specifically excite these resonant regions will be beneficial to assess the helical nature of the polymer. Reviews of these methods for quantifying polymer's helicity can be found in classical textbooks by Volkenshtein (1977) and Tinoco *et al.* (1978).

One method to take stock of the state of the electromagnetic field in the presence of a set of activities that the field will encounter is to represent the field in a tighter formulation. The entire field, with its two mutually perpendicular polarization directions, can be represented by a vector of two field components. The interacting elements can then be represented by a 2×2 matrix, sometimes called the Jones matrix. The process of propagating an EM field through these elements is then an operation by the element's matrix on the field. Thus the entire description of the field and all its sequential operations by material elements are relegated to matrix operations in this Jones Calculus (Shurcliff, 1962). However, the Jones method considers only pure (coherent) states. Another approach called the Stokes' vector is more useful for non-pure states. By using Mueller calculus, mixed (incoherent) ensemble averaged states can easily be represented. Because it is often more common to find large polymeric systems in a mixed state, the Stokes' representation is more useful. The main idea is that an electric field that is measured is derived from those elements of interaction had caused it to change its original state. An example may be a linearly polarized light encountering an optical quarter wave plate. This encountered object, the quarter-wave plate, may be represented by a matrix of transformation, converting the linearly polarized light to circularly polarized light. We can use the Stokes' vector to represent the observable in our matrix equation.

$$\mathbf{S} = \begin{pmatrix} s_0 \\ s_1 \\ s_2 \\ s_3 \end{pmatrix} = \begin{pmatrix} \langle E_x^* E_x + E_y^* E_y \rangle \\ \langle E_x^* E_x - E_y^* E_y \rangle \\ \langle E_x^* E_y + E_y^* E_x \rangle \\ \langle E_x^* E_y + E_y^* E_x \rangle \end{pmatrix}. \tag{2.34}$$

In this Stokes' vector, the relevant field intensities are quantified. Here, s_0 represent the total intensity of the light; s_1 represent the preference of the light source for 45° planar orientation; s_2 provides the preference for elliptical polarization; and s_3 measures the sense of ellipsometry. Thus, we can write an operator equation

$$\mathbf{S}_2 = \mathbf{M} \cdot \mathbf{S}_1 \tag{2.35}$$

where M is a 4×4 Mueller matrix representing the function (lens, polarizer, filter, rotator, polymer, any unknown medium). Successive passage of light through more operating elements can be written as

$$\mathbf{S}_n = \mathbf{M}_n \mathbf{M}_{n-1} \cdots \mathbf{M}_2 \mathbf{M}_1 \cdot \mathbf{S}_0 \tag{2.36}$$

where M_i represents the ith Mueller matrix of transformation encountered by the light that had its original Stokes' vector described by S_0. Thus, given knowledge of all but one of the Mueller matrices, a measurement of the Stokes' vectors will yield the values of the unknown matrix elements and provide information about the unknown object.

In the field of *optical ellipsometry*, the idea is to use the measured Stokes' vectors to deduce the characteristic Mueller matrices that led to the resulting vector (Azzam and Bashara, 1977). Materials science fields have been using this concept for many years. Biophysicists have also used this approach to examine complex biological systems by measuring the birefringent properties of the molecules (optical activity) or systems of macromolecules. Because of the complexity of biological systems, often more than one lightwave frequency is needed to ascertain the characteristic of the molecule. ORD and CD spectra are used to assist in characterizing the structural content of complex biological macromolecules.

2.4.4 *Attenuated total reflection (ATR) and totally internal reflection (TIR) spectroscopy*

When electromagnetic waves incident upon a dielectric surface at an oblique angle, θ_i, part of the energy is reflected, and part of it is transmitted. The transmitted part leaves the surface at an angle θ_t. The relationship between these two angles is given by Snell's Law

$$n_1 \sin \theta_i = n_2 \sin \theta_t . \tag{2.37}$$

Here, n_1 and n_2 are the real indices of refraction of the incident and transmitted medium. We assume that there is no absorption in the excitation frequency regime. Hence $\eta = n$. Now consider the imposed situation where we want the $\theta_t = 90°$ from the surface normal. That means no real transmission into the medium since the wave will have its propagation vector lying in the plane of the surface. Under this constraint, the above equation becomes

$$\sin \theta_i = \frac{n_2}{n_1} . \tag{2.38}$$

The necessary condition for satisfying this equation is that $n_1 > n_2$. This means for such a condition to occur, the material from the wave incident side must be denser than the material on the other side. Furthermore, the critical angle (θ_c) that this phenomenon first occurs is defined by the ratio of these two indices of refraction (**Figure 2.4**). If we are traversing crown glass and then entering into the water, $n_1 = 1.53$ and $n_2 = 1.33$. Accordingly, $\sin \theta_c = 0.8693$. Thus, $\theta_c \sim 60°$.

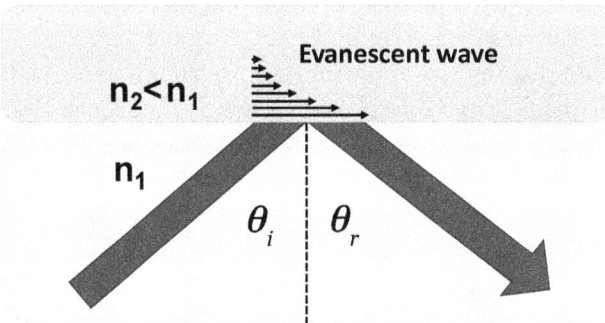

Figure 2.4. Evanescent wave develops when θ becomes $90°$. At that angle light θ_i is totally reflected at θ_r. The evanescent wave can interact with materials within the thin layer of surface.

By developing the electromagnetic field equations for the internal reflection condition described above, one finds that for all $\theta_i > \theta_c$, there is a non-propagating wave within the less dense medium. That is to say, the signal is not instantaneously zero, but that there is an exponentially damping signal. The damping length d is related to a damping coefficient κ in the form of $e^{-\kappa z}$.

$$d = \kappa^{-1} = \frac{\lambda}{2\pi\sqrt{\dfrac{\sin^2 \theta_i}{n^2} - 1}}. \qquad (2.39)$$

Here, λ is the wavelength of incident wavelength in vacuum and $n = n_1/n_2$. We see that if $\theta_i > \theta_c$, the damping length $d < \lambda$. What this result means is that if another dense dielectric material lies within that d distance, there will be some energy tunneled across to that medium. This idea of wave-tunneling, called *evanescent wave* coupling, can now be used advantageously in many optical experiments.

For example, if one is testing the reactivity of certain receptors for a strongly binding ligand with a very distinct absorption signal, e.g. C=O, we can deposit on the outer surface of an optical fiber the receptor molecules. As light is passing through the optical fiber, generally they will be totally internally reflected (TIR). Once ligands for this receptor binds to the receptor, the incident light encounters a new medium with a modified index of refraction at that interface if and when the light is at that wavelength corresponding to the C=O vibration frequency. Such phenomenon will lead to specific non-TIR condition for that wavelength, hence attenuated TIR, called *attenuated total reflection* or ATR, at that specific wavelength of absorptivity. This absorption signal shows only those ligands that are bound to the receptors, leaving the dominant solution ligands unseen by this probe. Thus TIR methods help to discriminate the adsorbed molecules from those that just simply exist in the solution and are further away from the surface. It has been pointed out by Zhu *et al.* (2007) that such a TIR configuration can be rendered highly sensitive to adsorbates quantitatively by measuring the polarization differential signal at these oblique angles. An oblique-incidence reflectivity difference (OI-RD) microscope has been shown to be sensitive for label-free high-throughput detection of biochemical reactions in a microarray format. Recent progress in this field allows the researchers to conduct real time, label-free measurements of reaction kinetics on sandwiched layers of adsorbates (Sun and Zhu, 2014).

There are several other ramifications or extensions to this technique, among them, the combination of TIR with Raman spectroscopy or with fluorescence will be discussed in Chapter 5. The topic of surface plasmon resonance is also gaining prominence in the biosensor literature.

2.4.5   Surface plasmon resonance (SPR)

The presence of an evanescent wave furthermore allows for coupling to surface plasmons in metal films deposited on a dielectric. Surface plasmons are surface electron waves that exist within metals. When the metal layer is very thin, only electrons propagating totally within the metal, hence parallel to the thin film, can exist. Thus, when a TM mode of optical wave enters the metal/dielectric (prism) interface, it alone has the capability of coupling with these electron waves, called surface plasmons. The maximum coupling occurs when the wavevector of the incident wave equals that of the wavevector of the plasmon wave (Homola, 2008). Since the surface plasmonic waves are propagating waves with a distinct propagating wavevector given by

$$\beta_{sp} = \frac{\omega}{c}\sqrt{\frac{\varepsilon_d \cdot \varepsilon_m}{\varepsilon_d + \varepsilon_m}} \tag{2.40}$$

where β_{sp} is the propagating surface plasmon wave of the magnetic field **H**, ε_d and ε_m are the complex dielectric constants of light of wavelength λ in the dielectric and metal, respectively, we simply match the two wavevectors. It should be stated that $2\pi/\lambda = \omega/c$. In order to have a propagating plasmon wave, for non-absorbing media, it is necessary that the condition of $\varepsilon'_m < -\varepsilon'_d$ is satisfied. For wavelengths in the visible and infrared region, the metals of Ag, Al, and Au are those that satisfy this necessary condition (**Figure 2.5**).

This matching condition is given by

$$\frac{2\pi}{\lambda}n_p \sin\theta_{sp} = \mathrm{Re}(\beta_{sp}) \tag{2.41}$$

where n_p is the prism index of refraction, and θ_{sp} is the angle of incidence within the prism that allows matching of the wavevectors within the plane of the metal film. When the matching condition is satisfied, we have

$$\theta_{sp} = \sin^{-1}\left(\frac{1}{n_p}\sqrt{\frac{\varepsilon_d \cdot \varepsilon_m}{\varepsilon_d + \varepsilon_m}}\right). \tag{2.42}$$

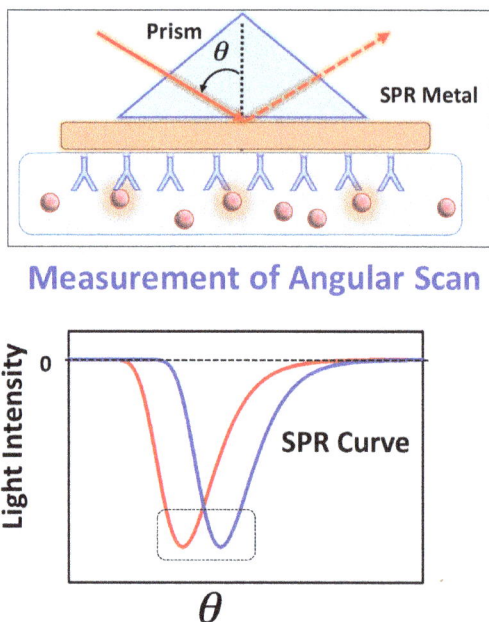

Figure 2.5. Surface plasmon resonance (SPR) principle. A prism is used to couple the photons by TIR to surface plasmons in SPR metal (gold). When tuned by changing the angle of incidence (θ) a resonance dip occurs. The dip changes (red to blue) with change in refractive index in the evanescent wave region.

Moreover, a strong dip occurs in the reflected optical signal, signifying maximum coupling between the plasmonic wave and the optical wave. Note that if the index of refraction of the medium is changed in any way, the resonance condition is no longer as strong, and the reflected signal rises. It is by such sensitive changes of the surface plasmonic conditions that this technique has become a much-considered one for probing the presence of minute or tracer quantities of matter that is adsorbed to the surface of the metal. Thus any analyte/solute/antibody that has an affinity to the receptor originally prepared on the metal surface will register a change in the local index of refraction. This change will cause a change in the θ_{sp}, the SPR resonant condition. Both time rate of change (sensogram) and the quantity of assay (angle scan) can be measured (Figure 2.5). Commercial surface plasmon resonance (SPR) instruments typically can detect *refractive index differential* (RID) at a level of 10^{-6}. Technology has advanced so that there are instruments using sophisticated dual wavelength incidence and wavefront cancellation methods that can eliminate common path effects to reduce noise. These methods are pushing the frontiers of detection to sense RID of 10^{-9} and beyond (Li *et al.*, 2008).

2.5   Two-photon Matter Interactions

2.5.1   *Scattering of light — Structural determination and diffraction*

In this section, we shall continue with our classical description of light interaction with matter. Whereas in the previous discussion, the key for detection of the presence and absence of matter lies mainly with the presence of some intrinsic signature, such as index of refraction change, absorption resonance, linear or circular polarization change, in this section, we consider the non-resonant interaction of light with matter.

Taking away resonance enhancement from our previous picture implies the signal now is much weaker than an absorption signal. However, we shall see that the process of "scattering" of light by matter does play a major role in providing information about these materials.

The physical idea of light scattering is as follows: imagine the atom as we had previously discussed, and once again light in the form of electromagnetic wave is impinging upon it. We had mentioned that electrons, being lighter than their nuclear counterparts, will respond to the classical fields with more displacement than the nuclei (the center of mass of the system does not shift much from the nucleus). Thus we consider a polarization field produced by this external field, **E** (**Figure 2.6**).

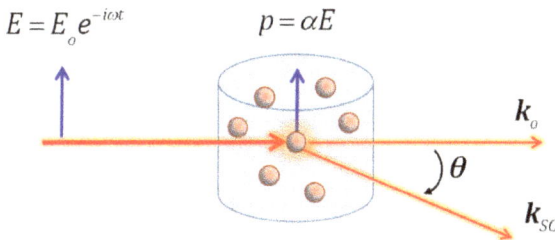

Figure 2.6.   Incident light of field strength E_0 and frequency ω enters material in direction $\mathbf{k}_0$, with many atoms, here shown in red. Each develops a polarization field, leading to dipole radiation. This scattering light is viewed at an angle θ. In the diagram, the scattering is viewed along the scattering vector direction $\mathbf{k}_{sc}$.

The induced dipole moment, $\mathbf{p}$ (Eq. (2.23)), will be given by $\mathbf{p} = \alpha\mathbf{E}$, where the polarizability, α, of the electron contributing to this signal is

$$\alpha = \frac{(e^2/m)\left[(\omega_0^2 - \omega^2) + i\omega\gamma\right]}{(\omega_0^2 - \omega^2)^2 + (\omega\gamma)^2} \approx \frac{(e^2/m)}{(\omega_0^2 - \omega^2)^2}. \qquad (2.43)$$

Note that since we are not near the resonance condition, the $\omega\gamma$ factor will be small compared to the first part of the denominator. Accordingly, the polarizability quantity can be approximated as shown in Eq. (2.44). Classically, a polarization field, no matter how small, will scatter light into all directions according to dipole radiation law:

$$E_{SC} = \frac{\ddot{p}_\perp}{c^2 R}. \qquad (2.44)$$

Here, the numerator essentially says that an accelerating electric dipole, from whatever the source, will emit light that is transverse to its mechanical degree of oscillation. This radiation then has, at far field point R, a $1/R$ radial dependence.

For a matter that has already an intrinsic polarizability, α_0, anything that adds to this total polarizability will be written as $\delta\alpha$.

$$p_i(\mathbf{r}, t) = (\alpha_{0i} + \delta\alpha_i(\mathbf{r}, t))E_0 e^{i(\mathbf{k}_0 \cdot \mathbf{r} - \omega t)}. \qquad (2.45)$$

In this expression, p_i means the polarizability of the ith matter particle under study. Furthermore, the imposing external field is given by

$$E(\mathbf{r}, t) = E_0 e^{i(\mathbf{k} \cdot \mathbf{r} - \omega t)}. \qquad (2.46)$$

Substituting these into the scattering field equation and now considering that we are collecting the scattering contribution from an ensemble of particles (assumed to be identical) over all space (either summing or integral), we have

$$E_s(R, t) = -\frac{E_0\omega^2}{c^2 R} e^{i(k_0 R - \omega t)} \int (\alpha_{0i} + \delta\alpha_i(\mathbf{r}, t))e^{-i(\mathbf{k}_s - \mathbf{k}_0) \cdot \mathbf{r}} d^3 r. \qquad (2.47)$$

Let us assume that the scattering atoms were such that $\alpha_{0i} = 0$, the remaining part of the integral becomes

$$\int \delta\alpha_i(\mathbf{r}, t)e^{-i\mathbf{q} \cdot \mathbf{r}} d^3 r. \qquad (2.48)$$

Here, the wavevectors, $\mathbf{k}_s$ and $\mathbf{k}_0$, are not collinear. So even if the scattering itself is an elastic process, $(\mathbf{k}_s - \mathbf{k}_0) = \mathbf{q}$, and

$$\delta\alpha(\mathbf{q}) = \int \delta\alpha_i(\mathbf{r}, t)e^{-i\mathbf{q} \cdot \mathbf{r}} d^3 r. \qquad (2.49)$$

The quantity $\mathbf{q}$ represents the Fourier difference wavevector between the incident and scattering optical wavevectors. Correspondingly, $\delta\alpha(\mathbf{q})$ is the qth spatial Fourier component of the fluctuations in the microscopic polarizability.

For elastic scattering, we now have the intensity of scattering as

$$I_s = \frac{|E_0|^2 \omega^4}{c^4 R^2} \left| \delta\alpha(\mathbf{q}, t) \right|^2 . \tag{2.50}$$

Note that since $\frac{\omega}{c} = k_0 = \frac{2\pi}{\lambda}$, the above expression shows that the scattering intensity is just proportional to λ^{-4}, the classical Rayleigh scattering limit. It is also of interest to note that the scattering intensity, I_s, is isotropic as long as $\delta\alpha(\mathbf{q}, t)$ is a constant. How does the polarizability fluctuation change? We look at the expression for $|\mathbf{q}| = q$:

$$q = 2 \left(\frac{2\pi}{\lambda} \sin\frac{\theta}{2} \right) \tag{2.51}$$

where q is the magnitude of the vector $\mathbf{q}$, λ is the wavelength of the light in the matter, and θ is the scattering angle. From this expression, one sees that an inverse relationship can be set up to describe the spatial extent of the Fourier vector with magnitude q, if we define $q = \frac{2\pi}{\Lambda}$, then Eq. (2.51) establishes a Bragg diffraction relationship:

$$\lambda = 2\Lambda \sin\frac{\theta}{2} . \tag{2.52}$$

Given a fixed radiation wavelength in this medium, λ, the "fluctuation wavelength," Λ, is inversely related to the scattering angle, θ. At the backscattering direction, $(\theta = 180°)$, $\Lambda = \frac{1}{2}\lambda$. This Λ is then the smallest length of fluctuation that will contribute to scattering signal at this wavelength, λ. For probing incident wavelengths in the 500 nm regime, fluctuations smaller than ~ 250 nm will not contribute significantly to the scattering signal. This also means that for particles of physical sizes much smaller than the wavelength of light, Rayleigh isotropic scattering will prevail. Such a result is consistent with the idea that the angular intensity pattern derived from light scattering in the Rayleigh–Gans or Mie scattering limit is due to particles of length dimensions larger than the wavelength of light. Prior to the advent of lasers, Peter Debye and his group had developed this technique to interrogate the structures and morphologies of polymers in the 1940 to 1950 era. Bruno Zimm (Zimm, 1948) extended this technique to develop the rigorous "Zimm plot" to characterize the changing of polymer size and shape, as measured by angular scans of the light scattering intensities.

2.5.2 Temporal studies: Dynamic Light Scattering (DLS) and Photon Correlation Spectroscopy (PCS)

From our expression for the scattered field, $\mathbf{E}_s(R,t)$ (Eq. (2.47)), we recall that the fluctuations in the local polarizability also has a temporal dependence: time t. This allows for the ability to carry out a different type of experiment and obtain dynamic data about the scattering particles in motion. The field of dynamic light scattering developed in the mid-1960s and has now claimed its position as one of the most useful methods for characterizing molecular size and shapes of dimensions much smaller than that which can be measured accurately using static (angular scanned) light scattering. The basis of this experiment lies with the expression

$$C(\tau) = \langle E_s(R,t) * E_s(R,t+\tau) \rangle \qquad (2.53)$$

where the expression $\langle \cdots \rangle$ corresponds to an ensemble averaged measurement of the field at time t *correlated* ($*$) to that of time $t+\tau$. The quantity $C(\tau)$ is the time correlation of the scattered electric fields. One sees that this is easily associated with the time correlation of the fluctuations in the local polarizability, $\delta\alpha(\mathbf{q},t)$:

$$C(\tau) = I_s = \frac{|E_0|^2\omega^4}{c^4 R^2} \langle \delta\alpha(\mathbf{q},t) * \delta\alpha(\mathbf{q},t+\tau) \rangle . \qquad (2.54)$$

So the time fluctuations in the polarizability will change the correlation function of the scattered light field. Dynamic light scattering (DLS) measures this fluctuating quantity, and then relate the measured correlation function $C(\tau)$ to the dynamic states of the particle system under interrogation. A fascinating aspect of this time correlation function is the connection of this quantity to the spectral function, $S(\omega)$. Scattering experiments historically had been conducted in the spectral domain as opposed to the time domain. These are related to one another through the Wiener–Kinchine Theorem:

$$S(\omega) = \frac{1}{2\pi} \int_{-\infty}^{+\infty} C(\tau)e^{-i\omega\tau}d\tau \qquad (2.55)$$

which basically says that the spectral function density is the temporal Fourier transform of the first-order time correlation function. In another form, if we define

$$S(\omega) = |E(R,\omega)|^2 \tag{2.56}$$

then,

$$|E(R,\omega)|^2 = \frac{|E_0|^2\omega^4}{c^4 R^2} \int_{-\infty}^{+\infty} \langle \delta\alpha(\mathbf{q},0) * \delta\alpha(\mathbf{q},\tau) \rangle e^{-i\omega\tau} d\tau \tag{2.57}$$

and the time correlation of the fluctuations of the **q**th spatial Fourier transformed coordinate is related to the spectral field function by a temporal Fourier transform (Chu, 1974; Berne and Pecora, 1976). The key element of DLS is identical to photon correlation function. So another name for this type of analysis is Photon Correlation Spectroscopy (PCS). From a historical perspective, DLS or PCS was also called Quasi-elastic Light Scattering (QELS) (Cummins *et al.*, 1964; Alpert *et al.*, 1965; Chu, 1991).

2.5.3 *Raman scattering (classical perspective)*

One of the most useful ways to look at Raman scattering is to examine the above-described process for a diatomic molecule. In this case, the polarizability fluctuations described above can have an added factor, namely the modulation of that polarizability by the vibration or rotation of the diatomic molecule. In the vibrational case, the classical description may be written as

$$\alpha(t) = \alpha_0 + \left(\frac{\partial\alpha}{\partial Q}\right)\delta Q(t) \tag{2.58}$$

with

$$\delta Q(t) = \delta Q_0 e^{i\Omega t}. \tag{2.59}$$

Here, the Q coordinate is the normal mode of vibration of the molecule in question, and the modulation index is basically the change in polarizability relative to the Q coordinate. The time dependence of the polarizability is simply the change due to the normal vibrational coordinate mode frequency. Based on this analysis, the usual way to obtain Raman data is to conduct a spectroscopic experiment, measuring the quantity $S(\omega)$ directly. From our analysis for the time-dependent polarization fluctuations, we can write

$$E(\omega) \propto \int_{-\infty}^{+\infty} \frac{\partial\alpha}{\partial q} \delta Q_0 e^{-i(\omega\pm\Omega)t} dt \tag{2.60}$$

$$S(\omega) = I_{\mathrm{S}}(\omega - \Omega) + I_{\mathrm{AS}}(\omega + \Omega) \tag{2.61}$$

where I_S is the spectral Raman signal for the Stokes component and I_{AS} is the Raman signal for the Anti-Stokes component. So essentially, the Raman signature provides a "fingerprint" of the particular molecular motion (vibrational, rotational) that modulated the incident light source frequency.

The classical picture does not discriminate between the strength of the Anti-Stokes against the Stokes component. However, by resorting to some elementary quantum mechanics, where we can describe the Raman signal as coming from quantized energy levels of a molecule (instead of frequency modulation), it can be shown that the Anti-Stokes Raman component intensity typically will be down by the Boltzmann occupation factor of $\exp[-(E_n/(k_B T))]$. To illustrate this point, in **Figure 2.7**, we consider the molecular system to have two vibrational states, $|1\rangle$ and $|0\rangle$. Here E_n is the energy level of state $|n\rangle$ of the molecular system. The energy difference between state $|1\rangle$ and $|0\rangle$ is then $\hbar\Omega = E_1 - E_0$. Thus in general, occupancy probability of the upper state $|1\rangle$ will be lower than state $|0\rangle$ by that Boltzmann factor, now $\exp[-(E_1 - E_0)/k_B T]$. From the illustration, the Anti-Stokes signal will be less intense than the Stokes signal. Indeed conventional Raman spectroscopy usually focuses on the Stokes shifted component. We shall return to these concepts later.

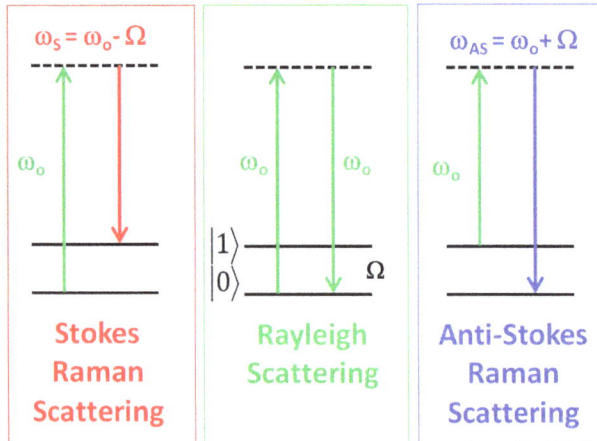

Figure 2.7. Rayleigh scattering is elastic with no frequency change. Scattering signal occurs at ω_0 (center). Stokes Raman signal shown as ω_S (red arrow) takes the excitation light ω_0 to a higher vibrational state $|1\rangle$ (left). In the Anti-Stokes scattering, the scattered signal returns at frequency ω_{AS} (right). Note it is at larger energy difference than ω_0 (blue arrow).

2.5.4 *Fluorescence spectroscopy*

Whenever we see the beauty of nature, the green of the grass or the rainbow set of colors of flowers, we are actually looking at the process of fluorescence in action. When light (usually over a wide spectral range) impinges upon a substance, grass blade or a flower or anything else, the light absorbed is converted for use by the substance that includes eventual emission of light at a different part of the optical spectrum. To explain this absorption, conversion and emission process, we need to resort to a simple little application of quantum mechanics.

Whereas macroscopic matter in its own right has a very small de Broglie wavelength, the individual atoms or small molecules do have de Broglie wavelengths that are much more comparable to the incident wavelength of light. As such, the interaction of light with individual molecules needs to be treated by quantum description. The most succinct description is the use of a *Jablonski diagram* and relate the processes to Einstein's A and B coefficients.

In **Figure 2.8**, the molecular energy states are assumed known and well-defined. We shall describe the process of calculating these states in another section. For the present, we have the S_i and T_i states called the electronic Singlet and Triplet states respectively. Here, i stands for the specific state identification. We have used $i = [0, 1]$ to denote the ground and excited states respectively in either manifold. The narrower spaced substates within each of the electronic states correspond to vibrational states of the molecule in question.

For any two states, e.g., S_0 and S_1, we now apply Einstein's expressions for light-matter interaction. Principally, with the atom or molecule in lower (or ground) state, incident light of energy density U at the appropriate energy $E = \hbar\omega = \hbar\omega_{10} = E_1 - E_0$, there will be an absorption rate, $B_{10}U(\omega)$, where B_{10} relates to the microscopic elements of energy absorption. According to the Einstein relationships, we can write

$$\frac{dN_1}{dt} = N_0 B_{01} U(\omega) - N_1 B_{10} U(\omega) - N_1 A_{10} \tag{2.62}$$

$$\frac{dN_0}{dt} = N_1 B_{10} U(\omega) - N_0 B_{01} U(\omega). \tag{2.63}$$

Once we assume the validity of the *microscopic reversibility condition* so that $B_{10} = B_{01} = B$, there is a full symmetry in the response of the individual states by the resonant incident radiation. However, the changes in

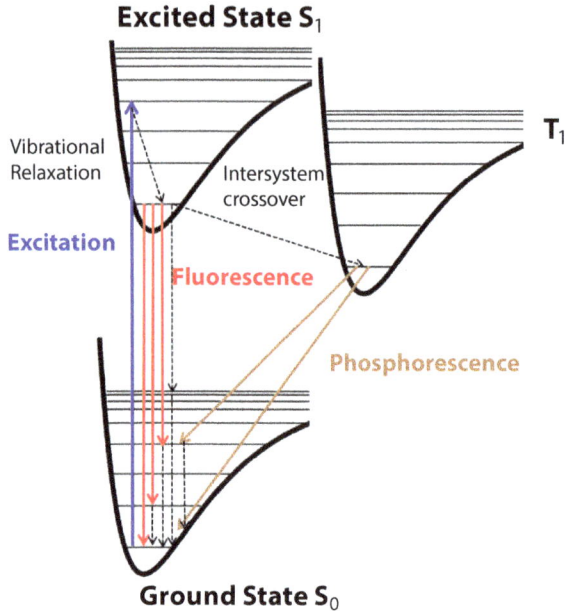

Figure 2.8.　Excitation (blue) at ω_0 takes electron into vibrational state manifold. From there vibrational relaxation drops the electrons into the lowest vibrational state rapidly. Fluorescence (red) radiation is one manner of de-excitation. Other means of excited state decay include non-radiative de-excitation (shown in dotted vertical line) and intersystem cross-over to the triplet excited state manifold, T_1. From the triplet state, electrons can return to ground state S_0 via spin-orbit coupling, a much slower process called phosphorescence.

the upper state have an added factor, called the *A-coefficient*, that describes the decay of the upper state via what is now called spontaneous decay. That is to say, whenever the system is not in the ground state, it will seek ways to return to the most stable ground state, including radiative decay through the quantum mechanically derived A-coefficient process. Assuming that the external radiation source has been turned off after a period of excitation, then the only non-vanishing part of these two equations is

$$\frac{dN_1}{dt} = -N_1 A_1 . \tag{2.64}$$

The solution to this simple equation is $N_1(t) = N_{10}e^{-A_1 t}$, where N_{10} is the initial excitation state population while $A_1 = k_f = \tau_f^{-1}$ the inverse of fluorescence lifetime τ_f. It is this process that we call molecular fluorescence.

It is important to note that how any excited states return to its ground state could reveal much about the structure of a molecule. Weber's group developed elegant means to measure fluorescence lifetimes (Weber, 1981) for characterizing molecular structure and dynamics. Let us take a closer look.

From the Jablonski diagram (Figure 2.8), we see several other features usually associated with fluorescence. First of all, we note that the broad, excited molecular vibrational state is not just a specific state that is decaying by a single wavelength fluorescence emission process. That is because the emission process usually is in the nanosecond regime, but the upper state has other more rapidly dissipating mechanisms. These include internal conversion to the lowest of the vibrational manifold of S_1, the possibility of converting to the ground state without any radiation via a non-radiative process, and in more complex systems, transferring the excited energy to the triplet manifold via the "intersystem (singlet to triplet) crossing" process. Each of these three processes has its own decay rate, but overall, they play a role in defining the efficiency of fluorescence process from this excited state.

The definition of the quantum efficiency ϕ of the process in question is

$$\phi = \frac{n_f}{n_i} \tag{2.65}$$

where n_i is the number of photons of energy $\hbar\omega$ incident onto the molecular system while n_f is the number of fluorescent photons present from this particular excitation. Obviously this efficiency is related to time constant for the various processes that are competing:

$$\tau_f = \frac{1}{(k_r + k_I)} = \frac{1}{k_f}. \tag{2.66}$$

So the time constant for fluorescence decay, τ_f, is related inversely to the sum of the rates of radiative decay, k_r, and non-radiative decays, k_I.

Thus the process of fluorescence is a three-step process, starting with the excitation from a lower or ground state to an upper state via an allowed electric dipole process. Upon absorbing the incident photon (or light), the system may or may not self-adjust to reach the lowest, metastable energy state that is capable of decaying toward the ground state. This is the second step of the process. Finally, in this final upper state, there is a lifetime or rate that governs the decay to the ground state, in a step that is totally not related to the initial state in any phase coherence sense. Only when the second step does not exist, do we have *resonance fluorescence*, and under that condition, phase coherence remains.

Some of the common features to be pointed out regarding fluorescence emission are:

(1) Both the absorption and emission processes are principally *electric dipole* (E1) processes. Hence they are dipole selective regarding the quantum mechanical energy selection rules, and these processes are associated with specific optical polarization orientations.

(2) All of the competing processes will deplete the upper state population and will decrease the efficiency of the fluorescence process. Those that have experienced the intersystem crossing will decay to ground state much slower than any other processes because the Triplet $\rightarrow$ Singlet decay needs a higher order spin-orbit coupling, which is usually only about $1/137\times$ the strength of purely Singlet transitions. The radiative process associated with this weaker and slower decay step is called *phosphorescence*.

(3) When the non-radiative decay pathways prevent any electrons from returning to its ground state radiatively, the system is considered *photobleached*.

(4) Since the environment of the fluorophore will affect the way a fluorophore configures, one can utilize the measured change in fluorescence time constant or efficiency to proclaim changes in the fluorophore's environment.

(5) Since the emission process is an E1 process, a measure of the changes in decay time as a function of the state of the polarization implies that the fluorophore has undergone motion such as rotation during the measured time. Using *fluorescence anisotropy decay* to measure fluorophore dynamics is a well-established method.

(6) Due to the dipole nature of the excited fluorophores, a specific dipole–dipole interaction allows for the transfer of energy from one species to another, without direct collisions. This is the process of Fluorescence Resonant Energy Transfer or FRET. The process was first discovered by Forster in 1948.

A thorough treatise on every aspect of fluorescence spectroscopy by Lakowicz (1999) is a must-read.

2.5.5 Fluorescence Resonant Energy Transfer (FRET) analysis

Because this process is one of an excited dipole interacting with another molecule a distance R away through *dipole–dipole interaction*, the process has a very strong spatial distance dependence:

$$k_T \propto \left(\frac{\kappa^2}{R^6}\right)\left(\frac{\phi_D}{\tau_D}\right) J. \tag{2.67}$$

(1) Here, k_T is the rate of the energy transfer process from the donor D to the acceptor dipole, A. Since the fluorescence process requires the initial excitation of the donor fluorophore, this rate is proportional to the efficiency, ϕ_D, and inversely proportional to the time constant of the donor lifetime, τ_D. The other factors of this equation reflect the transfer process.

(2) κ^2 is the angular orientation factor between the donor dipole and the acceptor dipole. If the dipoles are rigorously oriented parallel to each other, this factor is 1.0. If the dipoles are perpendicular to each other, then $\kappa^2 = 0$. In most cases, we assume that the dipoles are totally free to orient in any direction. So this factor reaches the averaged "free-rotator" value of 2/3.

(3) Because this is a dipole–dipole energy process, the distance between the two dipoles, R, comes in at $(R^{-3})^2 = R^{-6}$. So measurable FRET signal becomes a very sensitive measure of small molecular distances.

(4) J represents the quantum mechanical aspects of these two dipoles, in particular, how much energy overlap is there between the two dipoles in that excited state. Thus this J is called the overlap integral and can be calculated using quantum mechanical means.

Fluorescence Resonant Energy Transfer has established itself as one of the most powerful methods for probing details of macromolecular distances. This method has been called the molecular ruler approach of characterizing molecules in close proximity.

2.6 Nonlinear Optical Susceptibility

The response of matter to intense laser light differs from that of low-powered illumination. That is because the mechanism of response becomes higher ordered in terms of the change in the polarizability as it relates to the applied field, **E**.

Here we use the Yariv Chapter 21 discussion (Yariv, 1968). Consider the macroscopic polarization field, **P**; this quantity is a measure of the change in the electric polarization field in the presence of an external field, **E**. Whereas in the previous discussion, when linear response function alone was discussed, we considered only the linear dependence of polarization on the field, $\mathbf{P} = \varepsilon_0 \chi \mathbf{E}$ (Eq. (2.24)), we now include the successive higher orders of **E**-field response as well:

$$\mathbf{P} = \varepsilon_0 (\chi_L \mathbf{E} + \chi_2 \mathbf{E} \cdot \mathbf{E} + \chi_3 \mathbf{E} \cdot \mathbf{E} \cdot \mathbf{E} + \cdots) \tag{2.68}$$

where ε_0 is the electric permeability in vacuum, $\chi_L = \chi$ is the usual 3×3 tensor, and **E** is the applied external electric field. However, here, we see that the response **P**-field has the linear response only as its lowest order term, but can have higher order responses that are proportional to higher orders of the applied field **E**. The **P** derived from the higher order effects is called $\mathbf{P}_{NL}$, because the χ_{NL} quantity is now a more complex tensor instead of being a 3×3 tensor. When we introduce the total *polarization field* into Maxwell's equations for calculating a response to the applied field, **E**, we find that the material oscillators will respond along different field orientations, according to the strengths of the different non-zero elements of χ_{NL}. Thus, once the nonlinear polarization field components are measured, the new features of more complex coupling between field and matter can in principle be determined.

We shall focus on the *second-order* nonlinearity here. Consider the specific Maxwell equation:

$$\nabla \times \mathbf{H} = \frac{\partial}{\partial t} (\varepsilon_0 \mathbf{E} + \mathbf{P}). \tag{2.69}$$

Since $P = \varepsilon_0 \chi E + P_{NL}$, where $(P_{NL})_i = d_{ijk} E_j E_k$, the d_{ijk} elements are the nonlinear susceptibility components of interest. The wave equation describing this effect becomes

$$\nabla^2 \mathbf{E} = \varepsilon \frac{\partial^2 \mathbf{E}}{\partial t^2} + \frac{\partial^2 \mathbf{P}_{NL}}{\partial t^2} \tag{2.70}$$

for the non-magnetic material that we have set $\mu_0 = 1$. Furthermore, ε is the linear part of the dielectric permeability in the medium. In this manner, we note that the wave equation has an effective source term that is couched in the presence of the nonlinear polarization.

2.6.1 Second-order nonlinearities

For the general second-order nonlinear response, the above wave equation (Eq. (2.71)) can be solved subject to the approximation that the development of the nonlinear fields is a process that is slow compared to the incident field oscillation rate. Consistent with that approximation, and limiting ourselves to the simplest of the nonlinear contribution, second harmonic generation (SHG), we can show that

$$\frac{dE_{3j}}{dz} = -\frac{i\omega_3}{2}\sqrt{\frac{1}{\varepsilon}}\,d_{jik}E_{1i}E_{1k}e^{-i\Delta kz} \tag{2.71}$$

where indices $[1, 2, 3]$ represent the fields under consideration while the field polarization orientation directions are given by $[i, j, k] = [x, y, z]$, and we have chosen E_{3j} to represent the field at second harmonic frequency at polarization orientation j. The incident field is E_1, degenerate with E_2, and with polarization of i and/or k. We have also used

$$\Delta k = k_3^{(j)} - k_1^{(i)} - k_1^{(k)}. \tag{2.72}$$

This first-order differential equation can be solved using the initial condition that $E_{3j}(t = 0) = 0$. That is, the second harmonic field starts off being non-existent. The growth of this second harmonic field is given by

$$E_{3j}(L) = -\frac{i\omega}{2}\sqrt{\frac{1}{\varepsilon}}\,d_{113}E_{1j}E_{ik}\frac{e^{i\Delta kL} - 1}{i\Delta k} \tag{2.73}$$

and the intensity of the second harmonic field at distance L within the material will be

$$I(2\omega) = E_{3j}(L)E_{3j}^*(L) = \frac{1}{\varepsilon}\omega_3^2 d_{113}^2 E_{1j}^2 E_{1k}^2 L^2 \frac{\sin^2(\frac{\Delta kL}{2})}{(\frac{\Delta k}{2})^2}. \tag{2.74}$$

Note that this SHG intensity at frequency $\omega_3 = 2\omega$ will have its maximum when $\Delta kL = 0$. This is referred to as the phase matching condition $\Delta kL = 0$. This means for any material of finite (non-zero) spatial (L) content, $\Delta k = 0$ is necessary for *phase matching* condition to be met.

Furthermore, this second harmonic intensity is proportional to the square of the incident intensity along the i and k orientations. This result describes the simplest of the nonlinear response function. Schematically, the SHG process can be outlined in an energy diagram as well (**Figure 2.9**).

For the macroscopic material case, where L is a large, measurable quantity, the only way to satisfy the phase matching condition is what we have outlined above: rendering $\Delta k = 0$. This is not as easy as an automatically

SHG

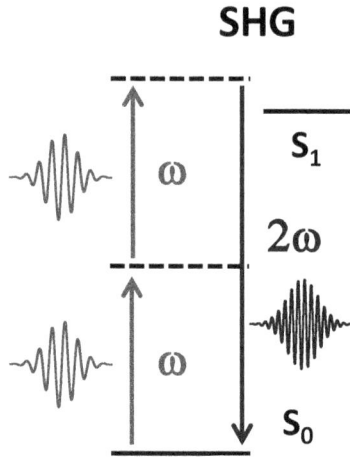

Figure 2.9.   Excitation of material at frequency $\omega = \omega_1 = \omega_2$ generates coherent emission at 2ω in this non-resonant SHG scheme.

satisfied situation because we encounter the necessity that $k_3 = 2k_1$. Since k_3 is the wavevector at twice the frequency of the excitation source at k_1 and ω_1, under normal wavelength dispersion conditions, this equality condition cannot generally be satisfied. So it takes macroscopic crystals of higher symmetry class than the cubic to provide the necessary variability in the indices of refraction to meet this criterion. Materials *without* a center of symmetry fall into the class subscribing to this needed requirement. For biological systems, two aspects help to ease this limitation: many molecular systems have asymmetries (e.g., helical) that are intrinsic to their structures, so they are intrinsically non-centrosymmetric. Furthermore, molecules being probed are often small in size, not macroscopic but are microscopic or nano-sized. This means the L quantity is small intrinsically, rendering the phase matching condition relatively easy to satisfy at any angle by virtue of the small distance, L. Thus $\Delta kL \sim 0$ is the effective phase matching condition. The downside of this is that all of the requirements for detectability relate to the physical magnitude of the nonlinear susceptibility elements. Because this is usually a small quantity, the signal is often very weak in biological systems.

   We further note from Figure 2.9 that the SHG signal is not necessarily associated with the transition into a real upper energy state, hence SHG does not normally depend on making two-photon transitions into S_1. However, resonantly enhanced SHG has been achieved, and signal enhancement can be, as expected, significant. It is interesting to note that this enhancement can be either at $\hbar\omega = E_1 - E_0$ or $2\hbar\omega = E_1 - E_0$.

The above discussion we have provided for SHG is equally valid for either sum frequency generation (SFG) or difference frequency generation (DFG) when ω_1 and ω_2 are not equal. Studies utilizing these nonlinear approaches will be discussed within the context of the biological systems of interest later in Chapter 6.

2.6.2 *Four-wave mixing processes*

The molecular systems (or crystalline systems) with the center of inversion will have third-order nonlinearity as its lowest order of nonlinear contribution. Following the path of our previous analysis, new phase matching conditions must be met with the criterion being

$$\Delta \mathbf{k} = \mathbf{k}_4 \pm \mathbf{k}_1 \pm \mathbf{k}_2 \pm \mathbf{k}_3 = 0 \qquad (2.75)$$

where $[1, 2, 3]$ signifies the three distinct (or degenerate) field components and "4" is the output signal with its wavevector k_4. The nonlinear susceptibility tensor for this general "four-wave mixing" situation is often very specific to the case under consideration and does not have the similar tensoral elements as we had displayed for the second-order nonlinearities. Some of the most commonly encountered third-order nonlinear effects in nonlinear optical applications are Stimulated Raman Scattering (SRS) and Coherent Anti-Stokes Raman Scattering (CARS). We show the energetic diagram of these two processes in **Figure 2.10**.

Consider first the stimulated Raman process, which is illustrated on the left side: We look at the schematic energy level diagram where the difference in optical frequencies of ω_0 and ω_s is just the energy difference of the two vibrational states, ν_1 and ν_0. That difference we call Ω. Thus

$$\omega_0 - \omega_s = \nu_1 - \nu_0 = \Omega. \qquad (2.76)$$

However, we can also write

$$\omega_0 = \omega_s + \Omega = \omega_s + \omega_0 - \omega_s \qquad (2.77)$$

which is a representation of four waves, two at ω_s, and two at ω_0. This representation suggests a degenerate four-wave mixing scheme. Indeed, the corresponding wavevectors will look like

$$\mathbf{k}_0 = \mathbf{k_s} + \mathbf{k}_0 - \mathbf{k}_s. \qquad (2.78)$$

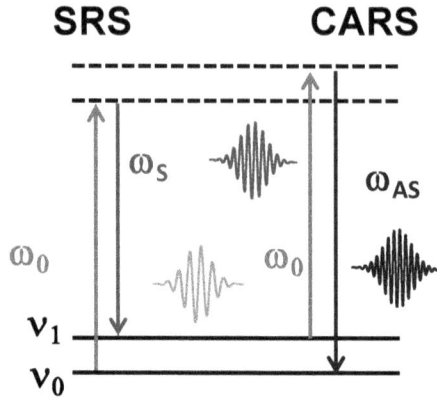

Figure 2.10.   SRS versus CARS. (a) SRS: since in spontaneous Raman scattering, the
Stokes component is given by $\omega_0 = \omega_S + \Omega$, leads to $\Omega = \omega_0 - \omega_S$. Thus the Stimulated
Stokes Raman signal is derived by the condition of $\omega_0 = \omega_S + \omega_0 - \omega_S$, leading to a
degenerate four-wave condition. (b) CARS: the four-wave mixing condition is reached
by $\omega_{AS} = 2\omega_0 - \omega_S$.

Thus the SRS signal is one that does not have strong requisite phase-
matching constraints, unlike the SHG signal. In fact, SRS should be considered a Raman gain amplifier. Usually, the backward scattered light encounters the maximum of a material capable of generating the Raman signal;
the gain is also highest in that direction.

Now we look at the Anti-Stokes side (right) of Figure 2.10. We see that
if we combine the incident wave twice with the Stokes wave, we shall be
able to arrive at

$$\omega_{AS} = 2\omega_0 - \omega_s \tag{2.79}$$

with the corresponding wavevector relationship given as

$$\mathbf{k}_{AS} = 2\mathbf{k}_0 - \mathbf{k}_s . \tag{2.80}$$

In such a scheme, the previously low occupancy upper level is now populated by the excitation of the Raman transition (shown on the right of
Figure 2.10). In this third-order nonlinear process, such an excitation is
taking place simultaneously with the introduction of the pump source once
more at ω_0, leading to the downward transition at ω_{AS}. The necessary
wavevector representation of this phase matching condition is shown in
Figure 2.11.

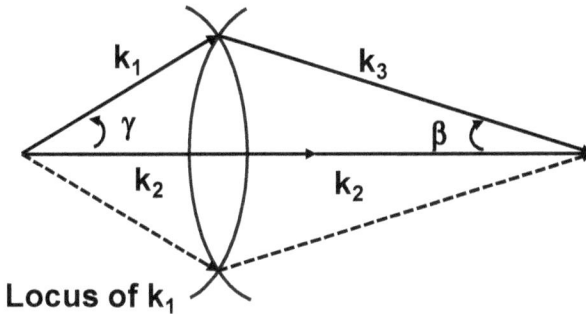

Figure 2.11. A construction for finding the loci of the direction of propagation $\mathbf{k}_3$ of the Anti-Stokes radiation.

It should be noted that the phase matching condition (Eq. (2.80)) is strictly required for bulk medium. So the CARS signal is observed at a specific angle satisfying the phase matching conditions in bulk nonlinear medium. For biological species of small molecular extent, with more relaxed phase matching conditions, CARS spectroscopy and microscopy have shown to be very useful as a technique for characterizing specific molecular vibrations.

Several other nonlinear optical phenomena of the third-order exist that can assist or become a detriment in biophotonics applications. One is *self-focusing* of the incident light, sometimes becoming so tight that an excessive amount of radiant flux can damage the region of interest. Another phenomenon is the *self-phase modulation* process. As the name sounds, this process can self-generate additional frequencies that may or may not be desirable in the specific application in biophotonics. We shall deal with these in specific encounters later.

2.7 Gradient Radiation Pressure ⇒ Optical Trapping

A revolutionary approach to biophysical research that developed in the 1990s is the use of optical trapping first discovered in 1986 (Ashkin *et al.*, 1986). An excellent review of the basics of optical trapping is given by Neuman and Block (2004). Essentially, the process taps into a well-known phenomenon of *radiation pressure*. A beam of light directed onto a surface and reflected from that surface will transfer momentum to the material, in order to conserve momentum. We normally do not feel the effect of that because the transferred momentum is smaller than that which is able to be sensed. Carrying this idea to smaller systems, consider a dielectric sphere of ~ 1 micrometer (μm) in diameter and a focused light beam onto that surface. The force felt by this sphere is given by

$$F_{\text{grad}} = \frac{1}{2}\alpha\nabla E^2 \qquad (2.81)$$

where α is the linear, non-resonant polarizability (scalar or tensor) of the sphere, and ∇E^2 is the *intensity gradient* of the imposing E-field on this sphere. What this means is that the larger the field intensity gradient, the larger is the force F_{grad}.

Figure 2.12 illustrates the principles of this phenomenon. In illustration (a), we show that with a Gaussian laser beam intensity profile, (higher intensity in more intense color), more momentum is imparted the dielectric sphere (bead) by the more intense part of the laser beam. For bead that is co-axial with the beam, the transfer of momentum after refraction by the dielectric sphere leads to a net forward force, providing what is normally called the "scattering force," however small this may be. In (b), should the particle be non-axial with respect to the incident laser beam, then the resultant net force will drive the particle toward the central axis of laser propagation. This is the essence of the *lateral force*, and even non-focused Gaussian beam profiles can move particles. Of course in a focusing laser system with a Gaussian intensity profile, this force is enhanced due to the

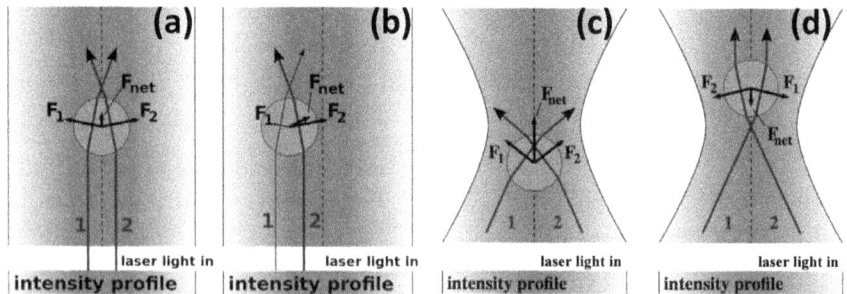

Figure 2.12.   Schematic illustration of the optical trapping of a dielectric sphere by laser beam with Gaussian intensity beam profile. (a) F_{net} experienced by the sphere that is co-axial with the laser beam is a small forward scattering force. (b) An off-axis sphere experiences net force F_{net} toward the center axis of the laser beam. This is the lateral force. In (c) and (d), a focusing Gaussian laser beam will further exert axial F_{net} to drive the sphere toward the center of the laser beam focus. Figure licensed under the Creative Commons Attribution 3.0 Unported License. Attribution: Roland Koebler.

increased intensity gradient. Hence most optical trapping configurations use such a configuration.

In illustrations (c) and (d), we show a focused laser beam near the "beam waist." Again due to the same gradient force transfer, a particle that is either in front of or behind the plane of the beam waist will experience a resultant force pulling it toward that equilibrium point. The combination of all of these four forces produces a trapping condition for the dielectric particle. A particle that has measurable dwell time within this focal condition is said to be "trapped." Such a particle achieves a stable position at nearly the center of the beam focus. Furthermore, the trapping strength is a function of the material polarizability, the field intensity gradient, and the numerical aperture of the focusing lens. Given that the non-resonant polarizability for dielectrics is not a large quantity, and a very strong laser beam can cause other deleterious effects on the focused object (i.e., heating), an essential element for efficient trapping is to have a large intensity gradient. This means in most cases trapping is achieved using the focusing lens of large numerical aperture (NA). For latex spheres of micron size, a 10 mW laser beam can be used in conjunction with a beam expander followed by a focusing lens with NA ~ 1.45 to achieve pico-Newtons levels of force. The *stiffness* of the laser trap is a function of the gradient intensity, and this quantity can be measured using the approximation that the force to remove the sphere from the center of the trap is of Hookian nature, $F = -kx$. Many methods exist to provide a measure of the stiffness of the optical trap. Among these are the modulation approach and the random displacement method.

Over the years, multiple-laser traps and single laser multiple traps have been developed. The advantage of the multiple-laser single trap is the ability to control the particle position within a much wider spatial domain than the focal region from a lens of NA ~ 1.45 or higher. This approach also lends itself to manipulating the beam position more controllably. The drawback of the multiple-laser single trap is the necessary alignment for creating the precise control of the trap. A single laser beam that can be diffracted or otherwise distributed into many point foci can be used to perform trapping on demand in specific locations in space. We shall encounter some of these unique configurations when we move into the specific needs of optical trapping and molecular force measurements in biophotonics investigations.

References

Alpert, S.S., Y. Yeh and E. Lipworth. Observation of time-dependent concentration fluctuations in a binary mixture near critical temperature using a He-Ne laser. *Phys. Rev. Lett.* 14(13): 486–488, 1965.

Arndt, M., O. Nairz, J. Vos-Andreae, C. Keller, G. van der Zouw and A. Zeilinger. Wave-particle duality of C-60 molecules. *Nature* 401(6754): 680–682, 1999.

Ashkin, A., J.M. Dziedzic, J.E. Bjorkholm and S. Chu. Observation of a single-beam gradient force optical trap for dielectric particles. *Opt. Lett.* 11(5): 288–290, 1986.

Atkins, P. and R. Friedman. *Molecular Quantum Mechanics.* Oxford University Press, 2011.

Azzam, R.M.A. and N.M. Bashara. *Ellipsometry and Polarized Light.* North-Holland Pub. Co., 1977.

Berne, B. and R. Pecora. *Dynamic Light Scattering: With Applications to Chemistry, Biology and Physics.* John Wiley & Sons, 1976.

Chu, B. *Laser Light Scattering.* Academic Press, 1974.

Chu, B. *Laser Light Scattering: Basic Principles and Practice.* Academic Press, 1991.

Cummins, H.Z., N. Knable and Y. Yeh. Observation of diffusion broadening of Rayleigh scattered light. *Phys. Rev. Lett.* 12(6): 150–153, 1964.

Homola, J. Surface plasmon resonance sensors for detection of chemical and biological species. *Chem. Rev.* 108(2): 462–493, 2008.

Lakowicz, J.R. *Principles of Fluorescence Spectroscopy*, 2nd edn. Kluwer Academic/Plenum, New York, 1999. Li, Y.C., Y.F. Chang, L.C. Su and C. Chou. Differential-phase surface plasmon resonance biosensor. *Anal. Chem.* 80(14): 5590–5595, 2008.

Loudon, R. *The Quantum Theory of Light.* Clarendon Press, Oxford, 1983.

Neuman, K.C. and S.M. Block. Optical trapping. *Rev. Sci. Instrum.* 75(9): 2787–2809, 2004.

Pedrotti, F.L.S.J., L.S. Pedrotti and L.M. Pedrotti. *Introduction to Optics.* Pearson Prentice Hall, Upper Saddle, NJ, 2007.

Shurcliff, W.A. *Polarized Light: Production and Use.* Harvard University Press, 1962.

Sun, Y.S. and X. Zhu. Real-time, label-free detection of biomolecular interactions in sandwich assays by the oblique-incidence reflectivity difference technique. *Sensors (Basel)* 14(12): 23307–23320, 2014.

Tinoco, I., K. Sauer Jr. and J.C. Wang. *Physical Chemistry: Principles and Applications in Biological Sciences.* Prentice Hall, Englewood Cliffs, NJ, 1978.

Volkenshtein, M.V. *Molecular Biophysics.* Academic Press, 1977.

Weber, G. Resolution of the fluorescence lifetimes in a heterogeneous system by phase and modulation measurements. *J. Phys. Chem.* 85(8): 949–953, 1981.

Yariv, A. *Quantum Electronics.* John Wiley & Sons, New York, 1968.

Zimm, B.H. The scattering of light and the radial distribution function of high polymer solutions. *J. Chem. Phys.* 16(12): 1093–1099, 1948.

Zhu, X., J.P. Landry, Y.-S. Sun, J.P. Gregg, K.S. Lam and X. Guo. Oblique-incidence reflectivity difference microscope for label-free high-throughput detection of biochemical reactions in a microarray format. *Applied Optics* 46(10): 1890–1895, 2007.

Chapter 3

Molecular and Cellular Structure

3.1 From Atoms to Molecules

Before we jump into the essence of biological molecules and learn how they carry out the necessary biological functions, we need to have some basic understanding of the structures of these molecules. This means we need to digress into a short discussion on the nature of molecular bonding and the different forces that are called into play to affect the stability or instability of a molecule.

3.1.1 Molecular bonding

Molecules are atoms that have exerted tight binding toward each other such that the atoms lose their individuality in favor of the collective. In the simplest molecule, the hydrogen molecule, the basic composition is that of two hydrogen atoms coming together. Since hydrogen atoms each have one electron in the $1s$ orbital, and the orbital allows two electrons of opposite spin quantum number to complete its shell, the two electrons of the hydrogen atom will share the common molecular orbital, but being Fermions with spin quantum number $s = 1/2$, they must be distinct from each other. Since $1s$ electrons are fundamentally both in the $n = 1$ electronic orbit, and $1s$ signifies that their orbital angular momenta are the same, each with angular momentum $l = 0$, the only distinction comes in the intrinsic electron spin orientation, m_s. Thus the two electrons occupying the lowest molecular orbital of the hydrogen molecule must have opposite spin orientations, $m_s = 1/2$ and $-1/2$. As Fermions, the total wavefunction for the electrons in this molecular orbital must be anti-symmetrical with respect to the exchange of the two electrons. If we write the atomic orbital wavefunctions of these two electrons as $\psi_A(1s)$ and $\psi_B(1s)$, then the two *orbital* wavefunction possibilities are

$$\psi_s = \frac{1}{\sqrt{2}}\left[\psi_A(1s) + \psi_B(1s)\right] \tag{3.1}$$

and

$$\psi_{AS} = \frac{1}{\sqrt{2}}\left[\psi_A(1s) - \psi_B(1s)\right]. \tag{3.2}$$

In order to be consistent with the Fermion properties of these electrons, the total wavefunctions (orbital · spin) must be anti-symmetrical with the exchange of the two electrons. Thus ψ_s orbital wavefunction is associated with the $S = 0$ state, the anti-symmetrical **singlet** spin state, while the ψ_{AS} orbital wavefunction is associated with the $S = 1$ spin state, with triplet multiplicity, symmetrical with respect to electron spin exchange. A most interesting observation is that the spatial distribution of the electron probability of these two orbital states differs significantly.

We note that (**Figure 3.1**) when these two $1s$ electrons are brought together, the anti-symmetrical wavefunction (right top) has a very low (zero) electron density in the mid-point between the protons of the hydrogen atoms. On the other hand, the symmetrical wavefunction (right bottom), shows high electron occupancy probability in that mid-plane. This symmetrical orbital state will be the stronger binding molecular orbital than the anti-symmetrical wavefunction state. Therefore, energetically, the ground state of this shared electron molecular orbital (MO) is the symmetrical orbital state, associated with the $S = 0$ spin state. This is the basis of the shared electron bonding theory or the basis of the general symmetrical

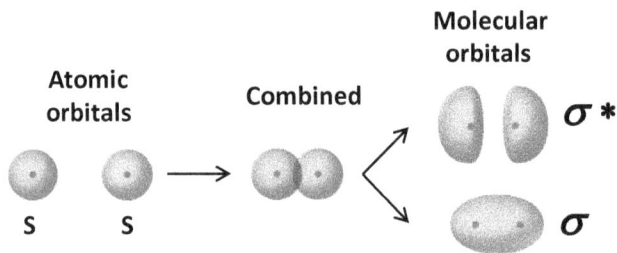

Figure 3.1. Formation of molecular orbitals. Two s orbitals combine to form the molecular orbitals, bonding orbital (σ) and antibonding orbital (σ^*). For example, the formation of two hydrogen atoms to form a H_2 molecule. Figure reproduced under Creative Commons Attribution 4.0 International License from OpenStax©.

orbital nature of most of the lowest ground states of covalently-bonded MO states. These states are given molecular orbital designations

$$\sigma(1s) = \psi_s = \frac{1}{\sqrt{2}}\left[\psi_A(1s) + \psi_B(1s)\right] \tag{3.3}$$

and

$$\sigma^*(1s) = \psi_{AS} = \frac{1}{\sqrt{2}}\left[\psi_A(1s) - \psi_B(1s)\right]. \tag{3.4}$$

The designation of σ and σ^* shows that these MOs are composed of the axially symmetrical atomic orbitals. Thus the binding will also exhibit axial or cylindrical symmetry with respect to the two nuclei.

Continuing with the idea of MOs being approximated by linear combination of atomic orbitals (LCAO), we can move toward the next higher energy atomic states (**Figure 3.2**). With principal quantum number $n = 2$, the possible atomic states are now $2s$, $2p_x$, $2p_y$, and $2p_z$. The wavefunction associated with the $2s$ orbital is still with angular momentum $l = 0$, hence spherically symmetrical. However, the $2p_{x,y,z}$ components will exhibit the anti-symmetrical distribution of the p-wavefunction, consistent with the $l = 1$ angular momentum state of the atoms. A comparison of these two states will show that when compared with their corresponding s-waves, their spatial extent is much larger, and their electron density $e|\psi|^2$ will have spe-

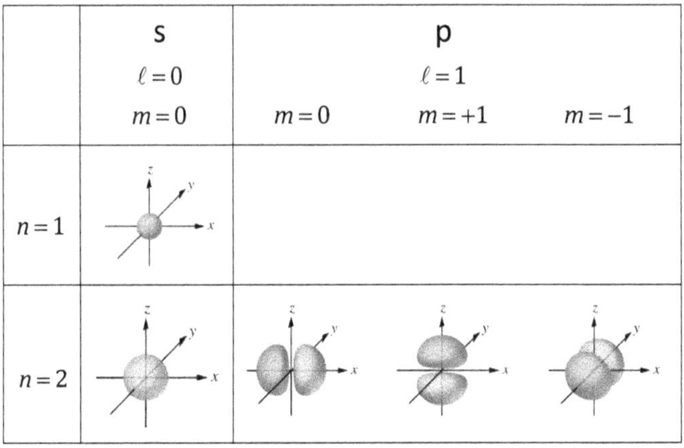

Figure 3.2. Shapes of atomic orbitals. Atomic orbitals s has $n = 1$, $l = 0$ and $m = 0$ while p orbitals have $n = 2$, with $l = 1$, $m = 0(p_x)$; $l = 1$, $m = +1(p_z)$ and $l = 1$, $m = -1(p_y)$. Figure reproduced under Creative Commons Attribution 4.0 International License from OpenStax©.

cific lobes extending into the x, y, z-directions. If the covalent bonding of two atoms were via the p-wavefunctions, then their spatial bonding distance will be much longer (**Figure 3.3**), but they are energetically weaker. Furthermore, they will have specific rotational asymmetry depending on the particular x, y, z-components.

Bringing these atomic orbitals together to form the MO, we will find that there are now MOs exhibiting these non-axially symmetrical features. Moreover, the p_z atomic orbitals can form π-MOs, all of which are energetically weaker than the σ-bonds (**Figure 3.4**). Note that the distribution of

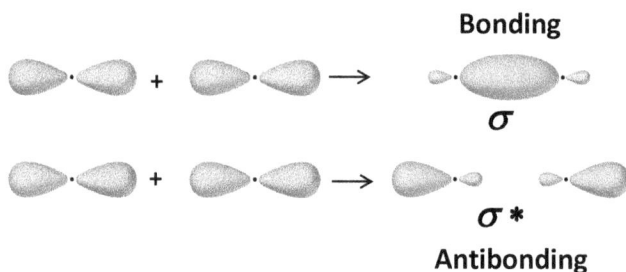

Figure 3.3.   Molecular orbitals involving p-orbitals. Two p atomic orbitals combine lengthwise to form the bonding (σ) and antibonding molecular orbitals (σ^*). Figure reproduced under Creative Commons Attribution 4.0 International License from OpenStax©.

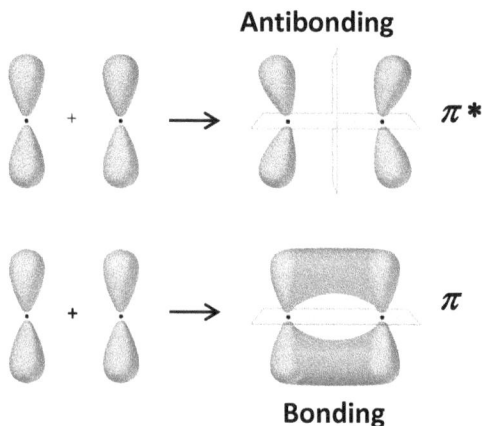

Figure 3.4.   Molecular orbitals involving p-orbitals. Two p atomic orbitals combine side-by-side to form the bonding (π) and antibonding molecular orbitals (π^*). Figure reproduced under Creative Commons Attribution 4.0 International License from OpenStax©.

electron density in the bonding π-orbitals have distinguishing $l = 1$ angular features. Specifically, these are the directed bonding orbitals associated with $2p$ atomic orbitals. They have the nomenclature *bonding and antibonding-MOs* (π and π^*) all of which are energetically weaker than the σ-bonds.

3.1.2 Hybridization and the unique carbon bonding capability

Jumping over the atoms of He, Li, Be, and B, we come to carbon (C). The carbon atom has 6 electrons. Assuming that the first two fill the lowest $1s$ atomic orbital, that leaves 4 electrons to move in the atomic orbitals called $2s$, $2p_x$, $2p_y$, and $2p_z$. The energies of these four states in isolation are comparable, except for detailed consideration of coupling between the orbit and the spin, or spin-orbit interaction (called *fine structure*). Thus, since the fine structure energetics is relatively small when comparing against orbital energies, any linear combination of the four atomic orbitals could suffice for the structure associated with a molecular orbital for atoms encountering a carbon atom. These orbitals can form linear combinations with one another depending on the energetics presented by the collection of atoms. Many molecules, including almost all of the organic molecules, have carbon as one of its principal components. We expect the structure of molecules with carbon to be highly variable due to the availability of *hybrid orbital* possibilities.

A simple hybridization of the $2s$ and $2p_x$ orbital may be written as

$$\psi(sp) = \frac{1}{\sqrt{2}} \left[\psi(2s) + \psi(2p) \right]. \tag{3.5}$$

In this linear combination, the $2s$ and any one of the $2p$ wavefunctions form a hybrid orbital that will exhibit the sum and difference of electron density (**Figure 3.5**).

Note that in this sp hybrid orbital, one of the lobes (here along the x-direction) has a much longer electron density extension, indicating its larger spatial electron distribution. Correspondingly, the opposite lobe has very little electron density. Hence, this orbital configuration will favor stronger binding toward one direction.

The molecule shown in **Figure 3.6** is C_2H_2 or acetylene. We note that because of the sp hybrid orbital, the strong σ-bond between two carbons is established. Furthermore, the additional two non-hybrid p-lobes, the p_y and p_z orbitals, jointly form two weaker, but more delocalized π-bonds.

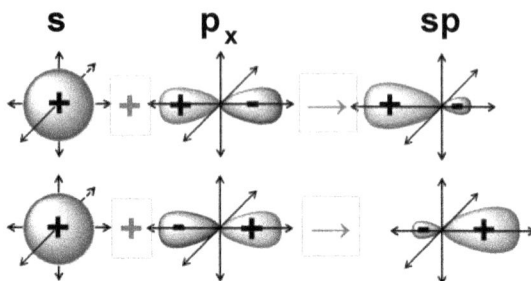

Figure 3.5. Formation of hybrid molecular orbitals. Atomic $2s$ orbital combines with one of the $2p$ orbitals (here p_x) to form a set of sp hybrid orbitals that point in opposite directions. Hybrid orbitals are spatially more extensive along one direction, hence stronger than molecular orbitals formed by p orbitals. The σ molecular bonds formed by p orbitals.

Jointly the C≡C bond is very strong, allowing the two extended sp orbitals to bind two H atoms to complete this molecule.

Similar to C_2H_2 (acetylene), C_2H_4 (ethylene), C_6H_6 (benzene), and methane CH_4 are varieties of carbon compounds that utilize the "hybrid orbital" constructs of $2s \pm 2p_x$ (sp hybrid), $2s \pm 2p_x \pm 2p_y$ ($sp2$ hybrid), and $2s \pm 2p_x \pm 2p_y \pm 2p_z$ ($sp3$ hybrid) to form molecules. In **Figure 3.7**, the ethylene molecule is a planar molecule with two carbons each in an $sp2$ hybrid orbital configuration, hence planar, to provide one σ bond between the carbons and four CH σ bonds. The remaining bond is a π bond as shown.

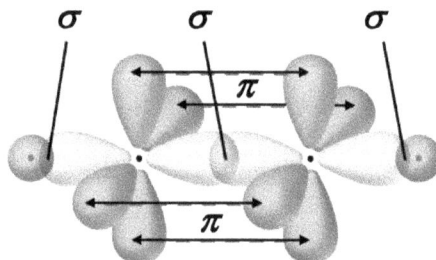

Figure 3.6. Molecular orbitals of acetylene C_2H_2. Molecular orbitals are formed between two sp hybrid C–H σ bonds and a C≡C triple bond involving one sp hybrid C–C σ bond and two C–C π bonds. Horizontal arrows show the connection of two lobes, indicate the side-by-side overlap of the four unhybridized p orbitals. The hydrogen atoms are shown at two ends of the molecule. Figure reproduced under Creative Commons Attribution 4.0 International License from OpenStax©.

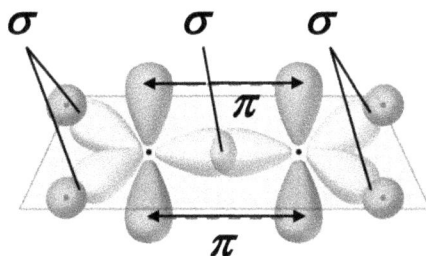

Figure 3.7. Hybrid orbital formation in ethylene. Each carbon atom undergoes sp hybridization by using a $2s$ and one $2p$ orbitals and the two carbon atoms form a σ (sp–sp in the center) bond with each other by using sp-orbitals. The $2p_y$ and $2p_z$ on each carbon atom which are perpendicular to the sp hybrid orbitals form two σ bonds between the two carbon atoms (shown by arrows). Figure reproduced under Creative Commons Attribution 4.0 International License from OpenStax©.

A methane molecule with its CH_4 constituents forms an energetically stable, tetrahedrally symmetrical structure, using its $sp3$ hybridization to space out the electronic orbitals in all four directions, each covalently bonded to a hydrogen atom (**Figure 3.8**).

The richness of carbonaceous compounds with all the variety of molecular structures has as its basis the capacity to form hybrid orbitals. Most compounds that form the core building blocks of the cellular machinery are long-chain polymers made up of carbon-based compounds using hybridization for molecular stability.

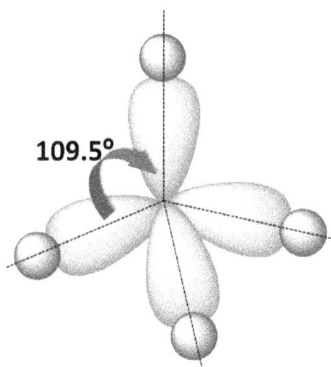

Figure 3.8. Hybridization of atomic orbitals in methane (CH_4). The carbon atom is $sp3$ hybridized with s-orbital and three p-orbitals ($2p_x, 2p_y, 2p_z$) of carbon atom. These $sp3$-hybrid orbitals are 109.5° apart. Each of the $sp3$ hybrid orbital joins a $1s$-orbital of H-atoms and forms four σ-bonds.

Besides the hydrocarbons mentioned earlier, several other elements play important roles in the structure of biological molecules. These are Nitrogen (N), Oxygen (O), Phosphorus (P), and Sulphur (S). Interestingly, they occupy the groups in the periodic table just *above* carbon, and they have 5 or 6 electrons in the outermost atomic orbital in their ground state. What that means is that they too have unfilled outer orbitals that can easily make covalent bonds with other atoms such as C and H, and they too can form hybrid orbitals to strengthen their bonds, rendering a variety of very well-known molecular groups in the biological world: H_2O (water), NH_2^+ (amine), COO^- (carboxyl), PO_4^{-3} (phosphate), and SH^- (sulfhydryl).

3.1.3 *Other molecular bonding mechanisms*

3.1.3.1 *Steric hindrance*

The structures we have developed that are based on atoms sharing electrons for their stable forms are called the *covalent* bonds. These are the strongest of all possible molecular bonds. As we extend from one small molecule to a succession of similar molecules, each using the same covalent bonding scheme, it is possible that we encounter a polymeric chain, such as that shown in **Figure 3.9**.

In such a chain, we notice that all of the C–C bonds are single bonds, and they show the directed, σ-bonding features. However, is each of the carbon atoms now capable of rotating the complete 360° without hindrance? The answer is "no". If one examines the individual H atom associated with this chain, and we recall that the sharing of the hydrogen electron with the C atom leaves the extremity of each of the CH group with a slight positive (proton) charge; hence, in rotating this system about the C–C axis, the regions of H:H closest to each other will experience the largest charge repulsion. Thus the C–C chain system will have a preference for the H groups to be oriented 90° from one C to the next C. This slight charge repulsion causes a preference of the polymer chain to be specifically oriented in its three-dimensional structure. These structures are the result of steric hindrance.

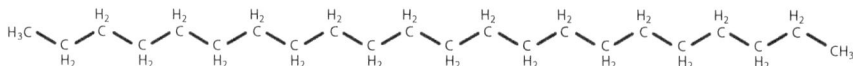

Figure 3.9. Polymeric chain of carbon with consecutive C–C bonds.

3.1.3.2 *Coulombic interaction*

Charge interaction is one of the major contributors of added polymer conformational constraints. Charge interaction is derived from Coulombic interactions and is given by

$$E_{\text{Coul}} = Z_A Z_B e^2 / \varepsilon r_{AB} \tag{3.6}$$

where the energy E_{Coul} is derived from two charges, $Z_A e$ and $Z_B e$ at a distance r_{AB} in a medium with a dielectric constant of ε. Clearly, the force is attractive if the charges, Z_A and Z_B, are opposite signs. This is particularly of interest if the group connected to the main carbon chain has an unshielded charge group, which we illustrate as follows: in this example, the CH_2 group has an ammonia group, NH_3^+, which allows it to interact with a carboxyl COO^- group (**Figure 3.10**).

Coulombic attractions are often as strong as the covalent bonds, e.g. Na^+Cl^- in the common table salt. Clearly such strengths, sometimes ~ 200 kcal/mol, can severely alter the molecular conformation of the polymer chain. However, it is seldom that biopolymers encounter bare ions. Since these ions are in an aqueous medium, there is much counter-ion shielding of these charges by the water and other ions in solution lying in-between the charged groups. These molecules function as a charge shield, typically extending to a *Debye shielding length* and can easily reduce the effective charges to just a fraction of the bare charge forces. In reality, this means that Z_A and Z_B are *effective* charges of ions A and B, each a small fraction of the bare charges of A and B due to this shielding. So in solutions, strengths of ionic interactions are only in the neighborhood of the strength of hydrogen bonds, which we will discuss next.

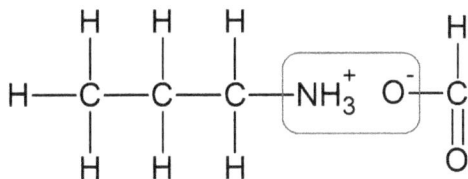

Figure 3.10.   Coulombic interaction between the positively charged ammonia (NH_3^+) and negatively charged carboxylic group (COO^-).

3.1.3.3 *Hydrogen bonding*

Water is as ubiquitous to biological systems as carbon compounds, and their presence in organisms may be as much as 80% by weight. Such high constituency implies their importance in life. Hence their molecular structure must be understood. Indeed, a water molecule in its own right is a very unique molecule. The oxygen atom has six electrons in its second $n = 2$ atomic shell. Thus it is necessary to distribute these six electrons within the four $sp3$ hybrid orbitals. Keeping in mind that we need to have the Pauli exclusion principle satisfied, the oxygen atom will have two of its orbitals filled with paired electrons, while the other two will have only one electron each. The unfilled orbitals can easily bind to hydrogen atoms to fulfill the molecular condition of H_2O. Since these levels are all nearly equi-potential, it is the local perturbations that allow for adjustment of which of the orbitals will be electron paired and which will have the H atom. In fact, the ability of the H atom to switch off from one oxygen orbital to that of another oxygen orbital with little or no energy penalty even at room temperature just due to thermal fluctuations is the source of hydrogen bonding. We show this in the following set of two diagrams (**Figure 3.11** and **Figure 3.12**).

First of all, we see that the lone-pair electrons create a net electronegativity for the oxygen atom. As a result, any small perturbation will render it preferable to attracting H atoms. This dipolar nature of water molecule and the fact that it is a very weak dipole allows for H atoms to jump from one oxygen atom to another.

The sharing of a hydrogen atom between two oxygen atoms is the essence of the water's hydrogen bond. The energetics of this sharing is

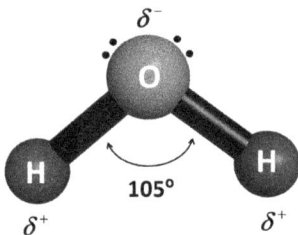

Figure 3.11. Molecular dipole moment of water. The asymmetry of the water molecule leads to a dipole moment in the symmetry plane pointed toward the more positive hydrogen atoms. Oxygen has a partial negative charge due to the lone electrons and each hydrogen has a partial positive charge, leading to a net dipole.

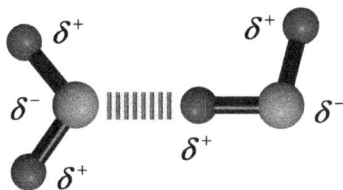

Figure 3.12. Hydrogen bonding between water molecules. Dipole–dipole interactions between the water molecules lead to hydrogen bond formation between the oxygen atom (partial negative charge, hydrogen bond donor) of one water molecule to one of the hydrogen atoms (partial positive charge, hydrogen bond acceptor) of another water molecule.

about 3.5 kcal/mol. Note that water in three-dimensional space will exist with a rather open network defined by numerous hydrogen bonds, as illustrated by the diagram (**Figure 3.13**). Due to the weakness of this bond, these bonds are of short time duration (nearly picoseconds). Thus many perturbations can and in fact do disturb the network. Such a fast exchanging network of hydrogen bonds in water is the reason for water to exhibit a rather high viscosity in its liquid state. This feature is especially useful as water molecules are in contact with a vast amount of organic molecules in the cellular environment. We will discuss this point later.

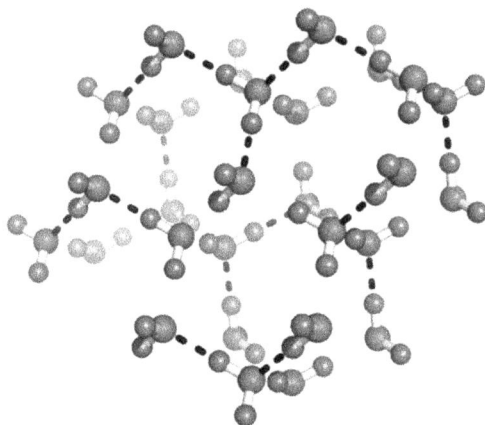

Figure 3.13. Hydrogen bonding between water molecules. Water molecules cooperatively form hydrogen bonds between them. The redistribution of electrons from the first hydrogen bond encourages formation of hydrogen bonds with other water molecules. Liquid water has the highest density of hydrogen bonding of any solvent and it has almost as many hydrogen bonds as there are covalent bonds. Figure made using PyMol (DeLano, 2002).

3.1.3.4   *Induced dipolar forces*

When molecules are in the presence of other molecules, even as charge-neutral entities, they will be able to polarize each other via the charge repulsion process and end up with induced dipole-induced dipole interaction. This is the basis of the dispersive force, which is commonly called the Van der Waals attractive interaction. Note that this term has a $-r^6$ dependence.

3.1.3.5   *Hardcore repulsion force*

When atoms and molecules are squeezed tightly against one another, they eventually run into the nuclear core, which is clearly highly charge repulsive. Thus for each atom of any molecular species, there is a distinct hardcore repulsive distance. This impenetrable distance for commonly encountered atoms of biological molecules is given in Table 3.1.

As seen in this table, these core distances are consistent with the increasing atomic number of the atoms, indicating the size is a function of Z.

Together with the attractive force described above, the resulting molecular interaction is often outlined by the total Van der Waals interaction shown (**Figure 3.14**). The Van der Waals energy is then the sum of the hardcore repulsion and the dipole–dipole attraction interactions. It is often given as

$$E_{vdw} = \frac{C_r}{r^{12}} - \frac{C_d}{r^6} \tag{3.7}$$

where the coefficients describe the hardcore repulsive and London dispersive forces involved. The impenetrable distance shown above is then the sum of the two atoms' Van der Waals radii. The dispersive or dipole–dipole

Table 3.1.   Typical impenetrable distances in molecules.

Atom	Distance range (Å)
Hydrogen	1.00–1.54
Oxygen	1.40–1.70
Nitrogen	1.55–1.60
Carbon	1.70–1.78
Sulfur	1.75–1.90

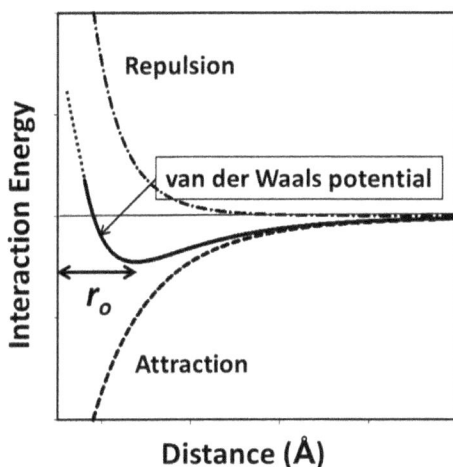

Figure 3.14. The Van der Waals potential. The combination of attractive forces (dashed) and repulsive forces (dash-dot-dash) lead to the net Van der Waals potential (black). The distance at which the energy minimum corresponds to the equilibrium distance for the interaction.

interaction has the familiar $-1/r^6$ attractive parameter. Energetically, the attractive part of this equation is often only ~ 0.5 kcal/mol. So it is an extremely weak interaction (Figure 3.14). In Eq. (3.7), $C_r =$ coefficient of the repulsive barrier while $C_d =$ coefficient of the Van der Waals attractive term.

3.1.3.6 *Hydrophobic interactions*

Up to now, the energies of interaction we have considered are what are considered "enthalpic," in thermodynamic notations. However, the minimum energy state speaks of the lowest *free energy state*, given by G (Gibbs) or F (or A for Helmholtz) free energies (at constant temperature and pressure, it is the G function; at constant temperature and volume, it is the Helmholtz free energy). Thermodynamics provides us with a method of counting energy such that accompanying the change in state, ΔF or ΔG, we have

$$\Delta F = \Delta U - T\Delta S \ \text{ or } \ \Delta G = \Delta H - T\Delta S. \tag{3.8}$$

In either case, the minimum free energy state is arrived by the combination of lowering the molecular energy of interaction, such as all of the different types we have just discussed, added to a contribution that is systemic,

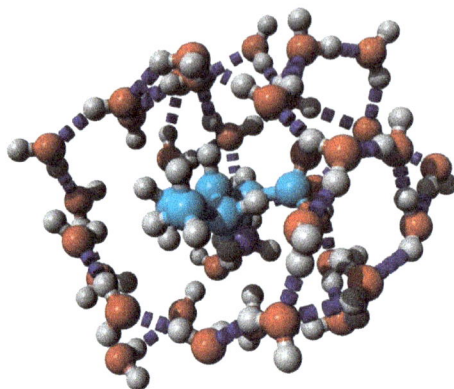

Figure 3.15. Clathrate. Representative example of a clathrate where the non-stoichiometric ratio of water surrounds the molecule. Figure made using PyMol (DeLano, 2002).

$-T\Delta S$. What this last term says is that for the system under consideration, the free energy minimum, ΔF, depends not only on the energetics of the system but also on the total environment of the system as far as its total *entropy* (or measure of disorder) is concerned. This is a very significant added consideration for molecules that survive principally in an aqueous medium. As we just discussed, water by itself exists in a rather open, tetrahedral structure, but only weakly bonded via the many hydrogen bonds. If, for example, a molecule such as a methane were to be inserted into its midst, and we know that methane is a hydrocarbon with no ability to hydrogen bond, first of all, the water to be displaced must sever a number of hydrogen bonds to accommodate the space needed for the methane molecule. This means the need to provide energy to break approximately four hydrogen bonds (HB), amounting to 1.2 kcal/mole. Upon insertion of the methane molecule, due to its tight CH_4 configuration that locks up all of the four carbon hybrid atomic orbitals, there will be the need to adjust the orientation of the water molecules surrounding this methane molecule. This accommodation will again cause disruption of the original pure water structure, removing a few HB but increasing its entropy, ΔS, to lower the total free energy.

As it turns out, methane clathrates do exist, but it takes many (upwards of 100) water molecules to rearrange to provide sufficient entropic contribution to counter the need to open up that water space. A representation of such a clathrate is shown in **Figure 3.15**.

The idea then is that *hydrophobic interaction*, in a collective fashion, does contribute to the overall stability of a molecule in water. The manner of decreasing effective free energy is to increase the entropy content of that system, allowing $-T\Delta S$ to lower the free energy in an overall sense. The magnitude of this collective contribution is usually in the neighborhood of a fraction of the strength of hydrogen bonds.

For reference of the above section, the textbooks by Atkins and Friedman (2011) and Tinoco *et al.* (1978) are excellent volumes to consult.

3.2 Molecules of the Cell

At the molecular level, a single cell looks extremely complex. Far from being a single, small molecule that we have been describing, these are very complex assemblies of molecules. On a cursory look, an optical microscope can visualize the presence of a cell as an irregularly-shaped body perhaps a few microns in diameter. It is compressible to the touch, but it is definitely elastic to a degree. The surface structure of such a cell may be smooth or very hairy, but it is well-defined. Upon more in-depth examination, a *eukaryotic cell* may possess some highly dense domains within the cell while other

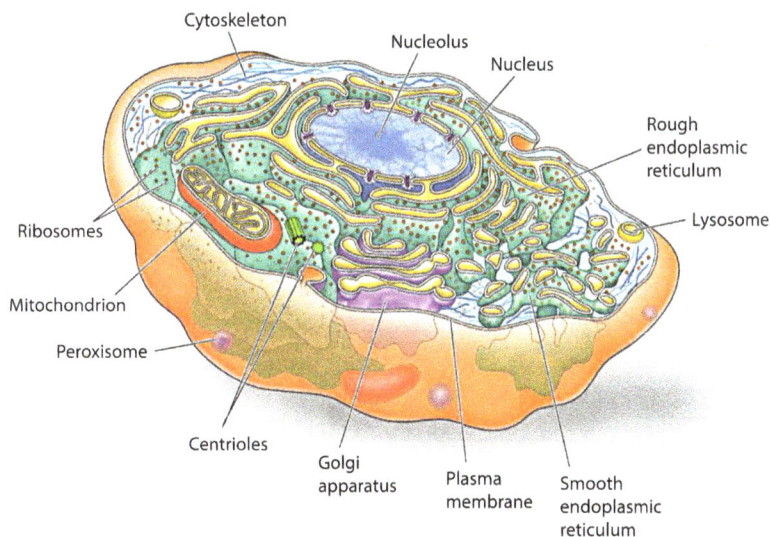

Figure 3.16. Representation of animal/plant cell. Plasma membrane, nucleus, cytoskeleton and many of the cytoplasmic organelles are marked. Figure reproduced with copyright permission from Cooper and Hausman (2007).

regions are less dense. A *prokaryotic cell* has a more uniform distribution of its material density. What makes these cells so well-defined? What are the dense domains within the cells? How do they survive in an aqueous, albeit nutritious and at the same time, hostile environment? The answer is complex, and we shall devote this short section to the structure and function of the cellular components (**Figure 3.16**).

3.2.1 *Nucleic acids*

The dense domains of the eukaryotic cells are the well-defined nucleus of the cells. That is not to say that the prokaryotic cells do not have nuclear matter, it is just that they are not so specifically confined. The key nuclear components are the nucleic acids deoxyribonucleic acid (DNA) and ribonucleic acid (RNA). The DNA molecule in its native state is a long-chained polymer.

3.2.1.1 *Deoxyribonucleic acid (DNA)*

One of the key components of the nucleus is the DNA molecule (**Figure 3.17**). Examining the detailed structure of this molecule, one notes that it is a composite of three molecular group elements: the phosphate group, the sugar group, which in this case is the deoxyribose, and one of the four nucleotides shown on the right, called the *bases*. Note that aside from the phosphate group, which we have mentioned before, the sugar group is a composite of COH in a five-membered ring assembly. This is sometimes called a carbohydrate entity, with just those three elements, C, O, and H. On the other hand, the nucleotides all have ring structures with N added into the mix. It is important to note that the nucleotides form two categories: purines and pyrimidines. Within each category, there are two specific molecules. The two purines associated with the deoxyribose are called Adenine (A) and Guanine (G), while the two pyrimidines are Cytosine (C) and Thymine (T) (**Figure 3.18**).

Interestingly, the single-stranded polymer of DNA is composed of chains of these molecules, where the chain connections extend from the C3′ and C5′ of each of the deoxyribose sugar groups. Thus a ssDNA is formed as shown in **Figure 3.19**.

Such a ssDNA molecule is highly flexible since the groups are linked by single σ bonds each with no constraints in rotation around the axis of binding. There are also no constraints as to how long such a polymer

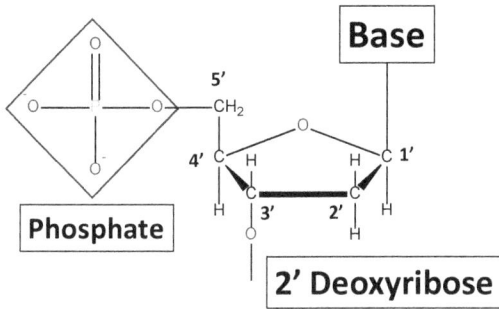

Figure 3.17. Structure of a nucleotide. Nucleotide contains three parts: pentose, a five carbon sugar, a base attached to the $1'$ carbon of the sugar and a phosphate group (diamond-shaped box) attached to the $5'$ carbon. The pentose here is defined as the $2'$ deoxyribose.

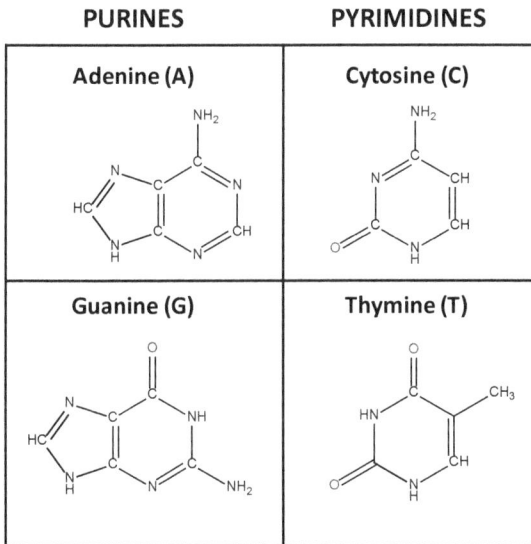

Figure 3.18. Chemical structures of the substituted purines and pyrimidines in DNA. Two purines, Adenine (A) and Guanine (G), and two pyrimidines, Cytosine (C) and Thymine (T) form the DNA bases.

can be. The key is that this chain contains only four possible pieces of variability: A, T, C, G, defined by the nucleic acid bases. In the study of the molecular biology of the gene, it turns out that such assemblies of DNA bases in specific sequence form a gene. As we shall see later, this important

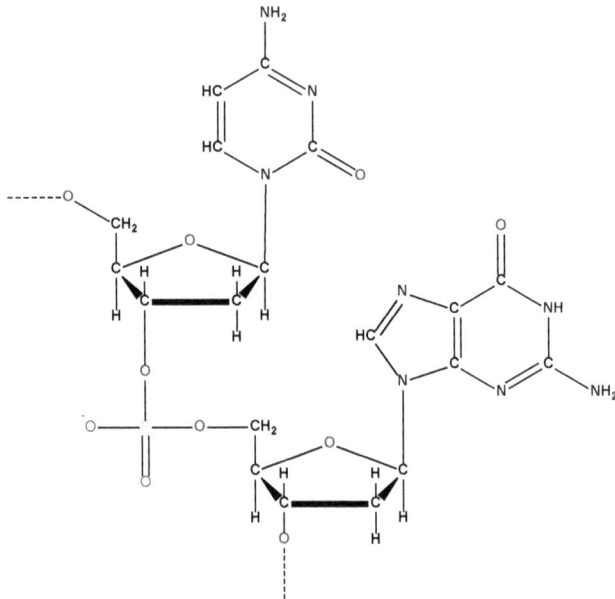

Figure 3.19.   Phosphodiester linkage in DNA. Two nucleotides (3′ end guanine and 5′ end cytosine) are linked by a phosphate group. The phosphate group and the two ester bonds form at the 5′ and 3′ carbons making the phosphodiester linkage.

recognition defines the modern molecular biology of the gene, and the DNA molecules in a unique sequence define genes specific for any organism as well as for cellular functioning.

The Watson–Crick model of the double helix provided additional insight as to how genes from one DNA strand can be transferred to another. The unique dsDNA has a pairing requirement such that only purine-pyrimidine pairings are possible. Thus a dsDNA can only have A–T or G–C pairings. This complementary pairing requirement allows for the information that is contained on one strand to be totally and faithfully transferred to the second strand!

In **Figure 3.20**, one sees that the information of the left strand is faithfully copied onto the right strand, but with complementary nucleotides. The actual structure of the B-form helix of the dsDNA is shown on the right. This conformation of the dsDNA is well-protected against anomalous sources that might damage the genes. The helical periodicity of the dsDNA in this B-form is 3.4 nm.

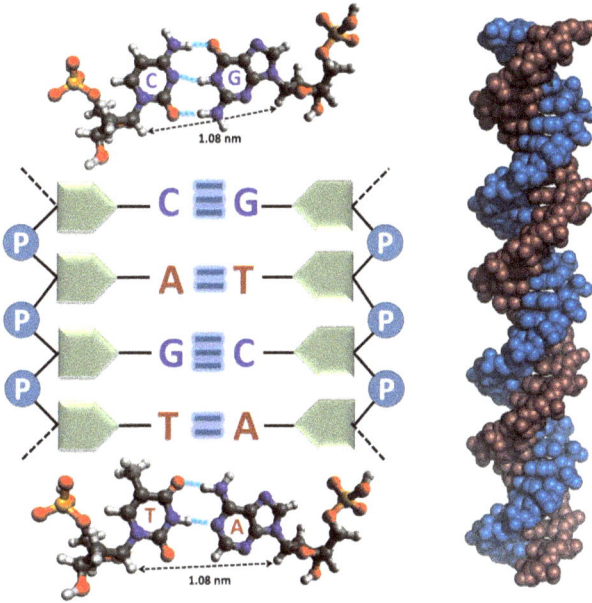

Figure 3.20. The Watson–Crick base pair formation. The phosphodiester linkage between sugar-phosphate chains are indicated in letters P linked by zig-zag lines on both sides. The left side chain is from the 3′ to 5′ end while the right side goes from the 5′ to 3′ end. The AT base pair has two hydrogen bonds (below) and the GC base pair has three hydrogen bonds (top) shown on the chemical structures. The distance between the 1′ sugar hydrogen atoms in both cases will be 1.08 Å. The double helical structure of DNA is shown on the right. The two strands 3′–5′ and 5′–3′ are shown in different shades. The figures are made using PyMol (DeLano, 2002).

The question that comes up next is: how does this information contained in the dsDNA get transcripted from mother to daughter genes, or translated into functioning proteins required for the survival of the species (cells)? An examination of the dsDNA structure shows that the double strand system is held together by hydrogen bonds between the purine and the complementary pyrimidine pairs (A–T and G–C). In order for the information that exists in duplicate fashion to be used, this set of hydrogen bonds must be broken and then a new complementary pair established. This energy expenditure is made via enzymes that can carry out specific functions on molecules. More specifically, the molecule that is needed to open up the dsDNA strand is called a "helicase," named so for its ability to open up the dsDNA helical structure. Upon opening, the ssDNA then needs to be properly paired to create the next stable duplex. A molecule that carries out

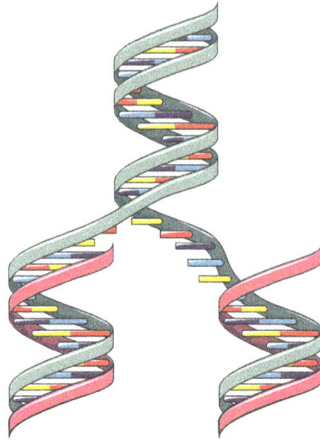

Figure 3.21. Semiconservative replication of DNA. The two strands of parental DNA separate and each will serve as a template for synthesis of a new daughter strand via complementary base pairing. Reproduced from Servier Medical Art licensed under a Creative Commons Attribution 3.0 Unported License.

that function is called a "polymerase." In the nuclear replication process, DNA polymerase is needed to recreate a daughter stable dsDNA strand (**Figure 3.21**).

We have yet to identify the molecular structures of these enzymes that carry out such a complex set of functions. However, for the moment, it suffices to say that these are *proteins* within the cells, and each plays a major role in the process of DNA replication.

3.2.1.2 *Ribonucleic acid (RNA)*

The fundamental structure of RNA is rather similar to the DNA except for several key distinctions: notice that in the C2′ position of the ribose group of RNA, a bonding group is a hydroxyl group (OH). In the DNA molecule, that same position is always occupied by a H. The non-existence of the oxygen atom defines one as the deoxyribose and the other one simply as the ribose sugar. Note that all other chain elements are identical, including the phosphodiester link that connects the C5′ of one ribose to the next at C3′ (**Figure 3.22**).

The other distinction between DNA and RNA is in one specific base. Whereas in the DNA structure, there is the pyrimidine called Thymine,

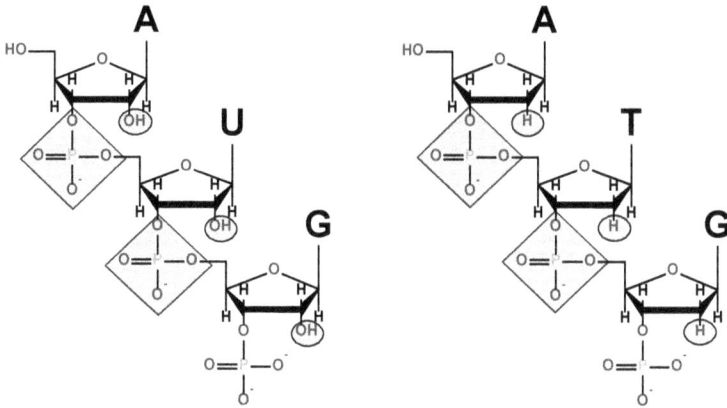

Figure 3.22. Phosphodiester linkages RNA and DNA. In both the RNA (left) and DNA (right) the nucleotides are linked by a phosphate group via the phosphodiester linkage. The differences between DNA and RNA are that the $-$H groups at the $2'$ position of the DNA are replaced by $-$OH in RNA and the base Thymine (T) in DNA is replaced by Uracil (U).

in the RNA, this is specifically replaced by Uracil (U). Thus the unique pairing in the RNA becomes A–U and G–C (**Figure 3.23**).

One asks: why is the RNA necessary at all if all information exists in the DNA, and we have the ability to duplicate (replicate) the DNA already? This is a good point, but the answer is that RNA carries out the other key function of the information center, that is to facilitate the translation of genetic information into the necessary proteins that can be used to carry out the functions of the cell. In order to do that job, there needs to be several new enzymes introduced: besides the necessary helicase, there now has to be an RNA polymerase to create a ssRNA from the information on a ssDNA. Such a piece of information then resides in the formed ssRNA, called an mRNA, for *messenger RNA*. That is, it carries the specific genetic information of need to be translated downstream into a particular protein (enzyme). This process of translation is another complex one. It takes place in a protein factory called the *ribosome* (**Figure 3.24**).

As schematically shown in Figure 3.24, as the ribosome passes the mRNA through its structure, a specific copy of three nucleotides is duplicated onto another RNA called the *transfer-RNA* (tRNA). The tRNA molecules then leave the ribosome and move into the cellular cytoplasm where the tRNA collects amino acids (to be discussed later) specific to the code called for by that tRNA.

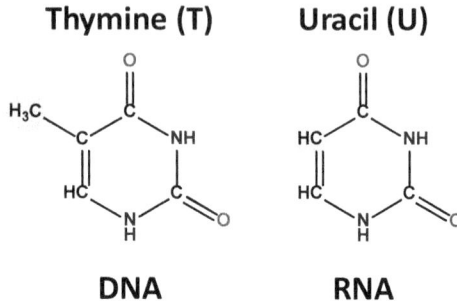

Figure 3.23. Chemical structural difference between Thymine (T) and Uracil (U). In RNA the purine base Thymine is replaced by Uracil (U). The methyl group in Thymine that is replaced by a hydrogen in Uracil is highlighted.

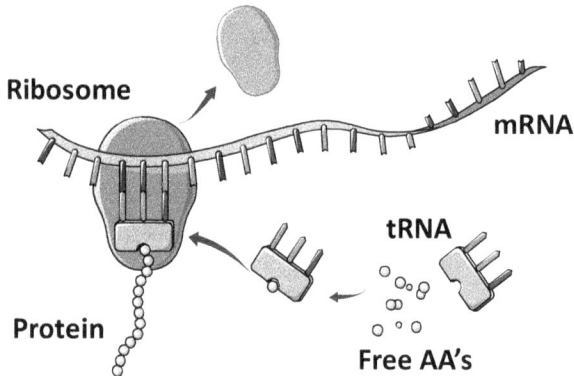

Figure 3.24. The protein synthesis machinery from transcription to translation. Transcription within the cell nucleus produces an mRNA molecule, which is modified and then sent into the cytoplasm for translation. The mRNA transcript is read by a functional complex consisting of ribosome and tRNA molecules. The tRNA molecules bring the appropriate amino acids in the sequence to the growing polypeptide chain by matching their anticodons with codons on the mRNA strand.

As prescribed, the coded tRNA (at the *anticodon loop*) (**Figure 3.25** and **Figure 3.26**) then singles out the specific amino acid molecule and binds to that one at the acceptor stem terminus. Upon binding to the specified amino acid, it is re-attracted to the ribosome, and within this assembly plant, the amino acids are linked together by elements of the *ribosomal RNA* (rRNA). The end product is a translated protein that fully reflects the specification of the original mRNA. This is the process of translation.

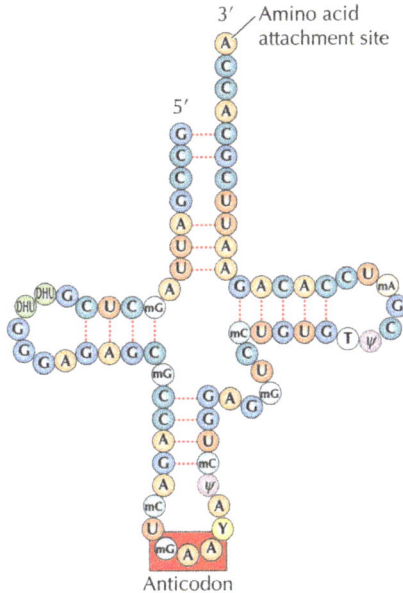

Figure 3.25. The structure of yeast phenylalanyl tRNA is illustrated in open "clover-leaf" form. Figure reproduced with copyright permission from Cooper and Hausman (2007).

3.2.2 Lipids

The fundamental requirement of the cell membrane is to contain the necessary biochemical elements within that confinement so that all metabolic processes of the cell are maintained. This is a tall order since the cell does not live in isolation behind closed walls. A cell must get nourishment from the environment, excrete waste products from within, and selectively determine if certain molecules should be allowed into the cell or certain molecules are desirable for making contact with the exterior of the membrane.

First of all, the environment that a cell lives in is an aqueous medium. That is, a cell is surrounded by water and molecules that reside in the solute state. This may include ions or other small molecules. The interior of the outer cellular membrane, called the plasma membrane, is the cytoplasm, and that is also mainly aqueous. In fact, about 75–80% of a cell's interior is water, H_2O. This implies that these membranes must maintain the proper solute equilibrium for all necessary components. In order to achieve that need, the membrane must be able (1) to shield exterior water from entering

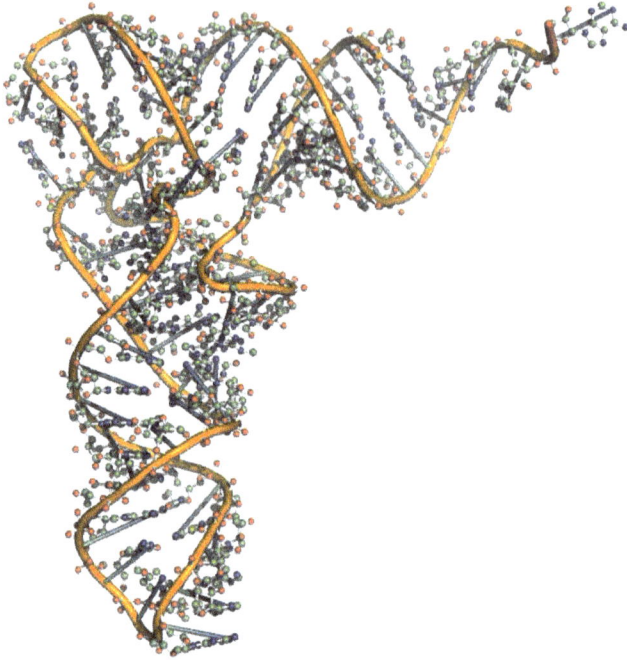

Figure 3.26. Three-dimensional structure of tRNA. Backbone and ball and stick model of the X-ray structure of yeast phenylalanine tRNA (pdb 1EHZ). Molecular structure produced using PyMol (DeLano, 2002).

into the cell, (2) create selective openings to admit needed molecules or ions on demand while rejecting other molecules and ions during the transportation process, and (3) define the parameters so that the exterior does not receive insults against the cell's well-being without being heavily challenged by defense mechanisms. These three criteria set the membrane up as a smart surface. Indeed, to satisfy condition (1), the membrane has as its primary constituents, lipid molecules that are set up as a lipid bilayer.

From the schematic diagram (**Figure 3.27**), one sees that the single lipid molecule has a hydrophilic head group and a long hydrophobic tail. The head group is usually some phosphor- glycerol- complex, allowing it to have a high affinity for water of the exterior environment. The tail group is nothing more than hydrocarbons in a long chain, either saturated or sometimes unsaturated as indicated, creating a strong hydrophobic domain. Given that molecular configuration, when in an aqueous medium, these molecules will attempt to achieve their lowest free energy configuration or

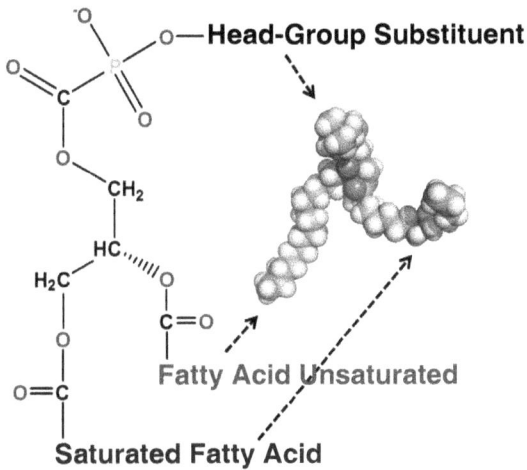

o—Head-Group Substituent

Fatty Acid Unsaturated

Saturated Fatty Acid

Figure 3.27. Schematic diagram of a phospholipid. Phospholipids contain a glycerol backbone linking two acetyl chains to head groups that are phosphoesters. The acyl chains vary in length and in the number of double bonds. The acyl chains have combinations of saturated or unsaturated fatty acids.

most stable configuration. That is to form micelles, lipid bilayers, or liposomes, as indicated. One notes that the membrane of the liposome indeed allows aqueous pockets within as well as in contact with an aqueous medium at the exterior. So a cell may be considered a very large liposome where within the interior resides the cytoplasm and the nucleus. Even for water molecules to penetrate such a membrane barrier, they must overcome the steep hydrophobic energy barrier set up by the double layer of hydrocarbon tails, arranged in the tail–tail configuration (**Figure 3.28**).

Endoplasmic Reticulum (ER) — From our very first figure illustrating the essence of a cell, it is also seen that the interior of the cell is very crowded with cytoplasmic substances. Nature has developed ways to put some networks of interior aqueous channels into this crowded environment. These then serve the purpose of allowing specific proteins and ions to travel within or along these channels at a much faster rate than otherwise possible due to restricted diffusion space. The structure of these channels is also mainly lipid molecules in bilayer construct. The *Golgi apparatus* is a well-known internal reservoir system.

In order to transport larger molecules from one ER component to another in an efficient manner, the cell allows for the formation of another membrane-based system: liposome (the small kind). These may simply

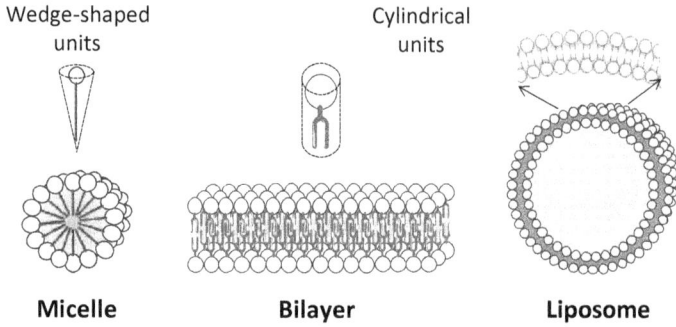

Figure 3.28. Phase separation of lipids in water. Lipids can be either single chain or double chain amphiphiles. Single chain lipids aggregate to form micelles and double chain lipids aggregate to form either planar bilayer structures or spherical bilayer structures (liposomes). Reproduced from Servier Medical Art licensed under a Creative Commons Attribution 3.0 Unported License.

be elements budding off from certain ER parts, carrying certain proteins needed at another location of the cell. Thus liposomes can be thought of as intracellular vesicles that can be used to transport substances within the cell.

For each of the vesicular bodies described, shielding against the undesired environment is one primary purpose. However, it is also necessary to transport elements from within to the exterior, and vice versa. This is our necessary criterion (2): the membrane must be able to sense the presence of required molecules or ions from exterior mediums and transport them to within (or without). There are basically two ways for this to be done: open aqueous pores to let in the selective ions or to allow molecules to diffuse in or out along concentration gradients, or to trigger an active transport process by which molecules or ions can travel into or out of a cell membrane against a concentration gradient. In the main, both of these processes will require the presence of membrane proteins to form a unique pore or active transporting structures. Illustrated in **Figure 3.29** are the two paths of ion transport. The path consistent with the concentration gradient is the one where open pores are formed, allowing passage of ions consistent with diffusion gradient. We illustrate both the Na^+ and K^+ ion pores allowing these ions to flow in their respectively gradient consistent direction. In the middle of this cartoon illustration, the active transport molecular motor is illustrated.

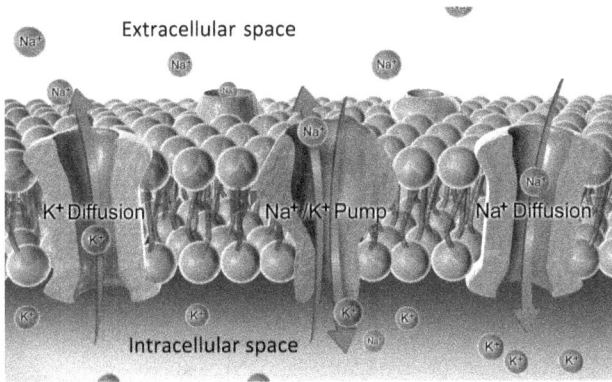

Figure 3.29.   Schematic representation of ion gradients and resting membrane potential. Figure reproduced with copyright permission from Staff (2014).

Active transport means that a network of molecules can do work to translocate an ion or molecule from one side of the plasma membrane to the other, even in the presence of high concentration gradients. Upon a command signal, this network of molecules will utilize the energy available via ATP hydrolysis to effect a change in molecular conformation to ease the translocation of the desired ion or molecule.

Note that in this schematic, the active transporter takes Na^+ ions from the intracellular domain to the cellular exterior and substitutes that with potassium ions (K^+) in reverse, both operating against the concentration gradient. Such a mechanism is necessary for controlling small amounts of ions needed for specific activity. Another illustration is the Ca^{++} pump in the sarcoplasmic reticulum, where Ca^{++} required for contraction process needs to be quickly removed after the required step is completed. In that case, the ions are pumped by the Ca^{++}-ATPase, another membrane-based motor.

Having mentioned ATP hydrolysis and its necessity in molecular motor functions, we note that one of the nucleotides that forms the energy storehouse for cells is *adenosine triphosphate*, ATP. In general, either Adenosine (A) or Guanine (G) can be the nucleotide, hence the general nomenclature of NTP. Energy is created by the breakdown of nucleoside triphosphate (NTP) into nucleoside diphosphate (NDP) or nucleoside monophosphate (NMP).

These processes are exothermic, generating the needed energy to cause changes in the local environment of molecules and thus molecular confor-

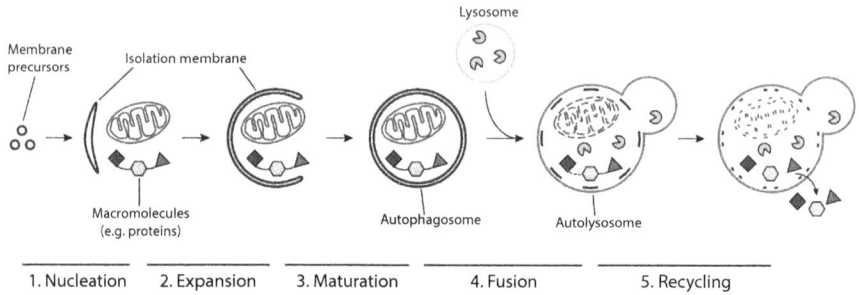

Figure 3.30. Schematic representation of autophagy delivering cytoplasmic material to the lysosomal compartment for degradation. (1) Nucleation: membrane donors nucleate an isolation membrane. (2) Expansion: the isolation membrane expands and engulfs cytoplasmic cargo material including organelles and macromolecules. (3) Maturation: the isolation membrane matures into a closed double-membrane autophagosome. (4) Fusion: the outer autophagosomal membrane fuses with a lysosome (or the vacuole in yeast), leading to the degradation of the inner membrane and the cargo. (5) Recycling: components are recycled back into the cytoplasm. Reproduced under Creative Commons Attribution License (Zaffagnini and Martens, 2016).

mations that lead to measurable work.

$$\mathrm{NTP} \rightarrow \mathrm{NDP} + P_i + 7.4 \text{ kcal/mol} . \qquad (3.9)$$

Thus the NTP is the energy source for molecular motors within the cell. Where are these NTP manufactured? They are made in another molecular factory that is called the *mitochondrion*. The mitochondrion is a complex organelle that lives within the cells of eukaryotic organisms. Within this complex double-membrane wall of the mitochondrion, there is another complex protein network that drives the formation of NTP, to be expended by the cell for its working functions.

The excretion of waste products from the cellular interior is another necessity of a viable cell. Often this process is done by the process of sub-vesicular formation (bubbles) where the unwanted constituents are encapsulated and expelled by vesicular coordination with the membrane, leading to the process of exocytosis. *Autophagy* (described more fully in **Figure 3.30**) is one of these processes whereby molecular elements captured by vesicles within the cell are expelled from the cell's interior.

Note that the driving force for fragmenting the intracellular components are once again enzymes (proteins) called protease or nuclease that reside within the autophagosomes.

Criterion (3) usually requires signaling receptors to be on a membrane's surface. The receptors are often COH groups in the form of long carbohydrate chains. To achieve selectivity for a particular molecule's signal transduction, the receptor often operates in cooperative or allosteric fashion. In that case, several receptors must operate in the capturing of a single signaling source to trigger the downstream activities. In order to collect these receptors and render them cooperative, it is often the responsibility of unique membrane complexes that drive the formation of these unique lipoprotein ensembles. Once more, it is clear, there is the need of having proteins to carry out the complex set of reactions downstream.

3.2.3 *Proteins*

In our previous discussions on genes and cellular integrity, we have frequently referred to the need for another type of molecule to provide the necessary specificity to biological function. DNA helicase, RNA polymerase, ATP synthetase are some of the elements of the cell that we have yet to describe in any detail. These are all parts of the protein system, often called the "workhorse" of the cellular machinery. These important molecules are composed of yet another arrangement of the atoms we had mentioned: C, O, H, N, and S. The primary idea behind this type of molecules is that there is, first of all, a similar backbone structure. Secondarily, the side groups, called residues, provide the amino acid specificity. Finally, the conformation of the polymer that we call "protein" depends very much on how the residues interact with its nearby environment, including water and ions. Thus, the functions of these proteins can be adaptable and specific.

3.2.3.1 *Primary structure of proteins*

The fundamental backbone structure of the protein is the basis of the amino acid. The molecule has an amine group (NH_2) on one end and a carboxyl group (COOH) on the other end. From the central carbon, there is a position to link onto a side group or residue (R). The specificity of the amino acid is totally imparted by the identity of the side group, of which there are 20. The ease by which amino acids can link to one another via backbone bonding is the acidic nature of these molecules (**Figure 3.31**). The bond that is formed is called an *amide bond* or peptide bond.

Again, due to the principal σ bonding nature, the C'–N bond has azimuthal symmetry relative to the bonding axis. However, the presence of the C=O allows a significant level of π-bonding, thus locking the group, NHCO into a planar structure (**Figure 3.32**). Indeed this is the base structure of the peptide bond from which variations are introduced with the presence of residues on both sides. Note that the C_α–C' bond or the C_α–N by itself has rotational symmetry about its axis. However, due to the presence of the different residues, many of which are rather large in size, there will be a steric hindrance (**Figure 3.33**), resulting in the NHCO forming a peptide plane.

The structure of the peptide or protein is then totally dependent upon the identity of the many residues, R, associated with each of these amino acids. A table provides the listing of the 20 commonly found amino acid residues (**Figure 3.34**).

AA Side Chain

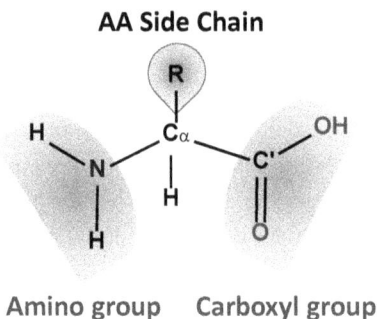

Amino group   Carboxyl group

Figure 3.31. Schematic representation of an amino acid. Amino group ($-NH_2$) and carboxyl group ($-COOH$) are attached to the central chiral carbon. The central carbon differentiates various amino acids with the side chain substitution.

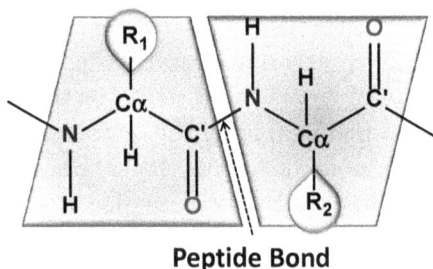

Peptide Bond

Figure 3.32. Schematic representation of the peptide bond between amino acids. The junction between two amino acids that reside in a protein is formed by the C=O group of R_1 (left, N-terminal) and N–H group of R_2 (right, C-terminal) residues.

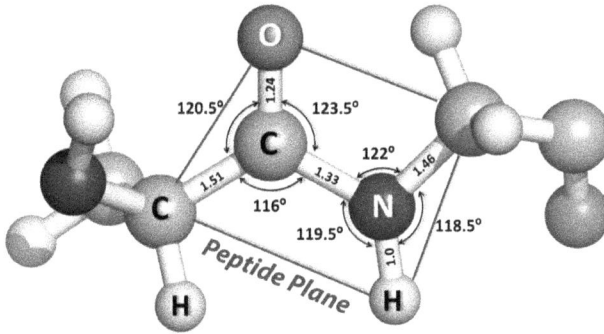

Figure 3.33. Geometry of the peptide plane: the structure of the peptide group along with various bond lengths and bond angles are shown. The amide plane (peptide plane) is restricted to planarity because of the partial double bond character of the C–N bond. Figure generated using PyMol (DeLano, 2002).

What we see here is that the side groups divide the different amino acids into three broad categories: nonpolar (hydrophobic), polar (hydrophilic), and charged (hydrophilic). Immediately it is evident that the role of water is very important in the formation of the conformation of the protein molecule. Let us take a closer look at the peptides as they build up into functioning proteins and enzymes.

In the buildup of this polymer chain, each peptide bond forms a rigid planar structure. The identity of a particular protein or enzyme is then defined by its **primary structure**, which is just the composition of the peptide chain, composed of many amino acids (**Figure 3.35**). Note that the rotational flexibility comes about solely from the C_α via the C_α–C' and the C_α–N bonds. The level of rotational degree of freedom is then a function of the residue and its environment. The angles that are defined are called the ψ and ϕ angles.

The averaged resulting angles in a ϕ–ψ plot define the protein's molecular conformation or secondary structures of the protein (**Figure 3.36**).

3.2.3.2 *Secondary structure*

The ability of the peptide to be flexible depending on the nature of the residues in their natural sequence allows for the formation of rather well-defined **secondary structures** of the protein. One of the most common secondary structure is the helical structure.

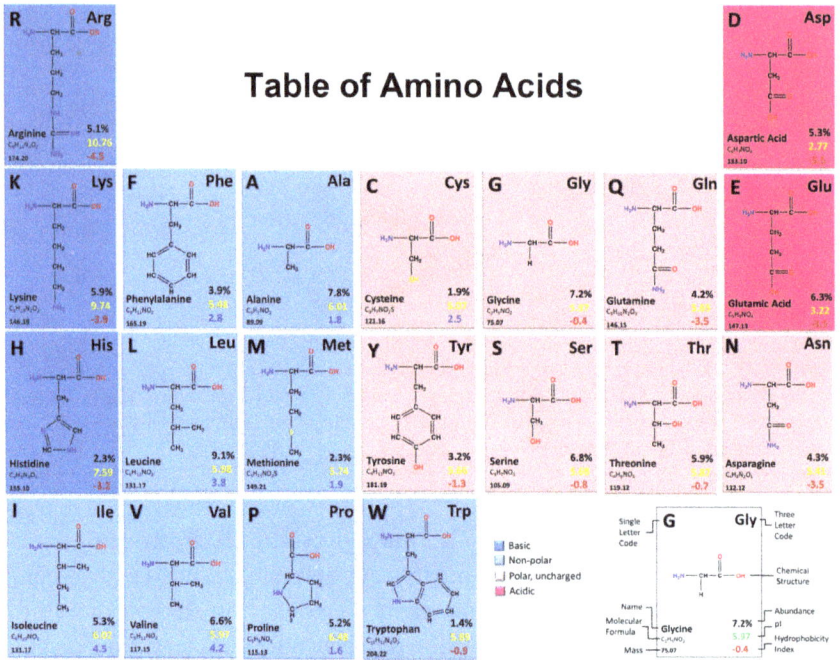

Figure 3.34. Amino acids. Table of 20 naturally-occurring amino acids are grouped into basic (dark blue), non-polar (light blue), polar, uncharged (light pink), and acidic (dark pink) panels. Each panel is identified with the corresponding single letter and three letter codes along with their respective chemical structure, full name, molecular formula, relative abundance, pI and hydropbobicity index.

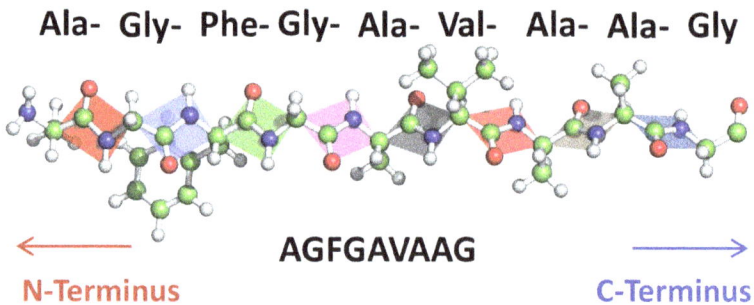

Figure 3.35. Molecular organization of the protein backbone. The molecular structure of a portion of a protein along with their amino acid sequence is shown. Each peptide plane is represented by various color plans along with the central swivel points between them that define the various conformations. Figure generated using PyMol (DeLano, 2002).

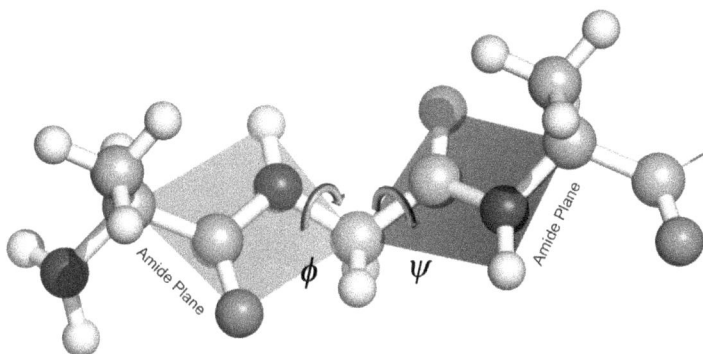

Figure 3.36. Definition of protein backbone torsion angles: the torsion angle ϕ corresponds to rotation about the N–Cα bond. The angle ψ corresponds to rotation about the Cα–C bond. The amide planes are also shown. The structure shown has $\phi, \psi = 180$ that corresponds to the C atom of the carbonyl group *trans* with respect to the C atom of the previous residue. Figure generated using PyMol (DeLano, 2002).

Amongst the possible helical structures, the α-*helical* structure is one of the most prevalent. This is a right-handed helix with a pitch of 0.54 nm and 3.6 nm residues per turn. The helix is stabilized by H-bonds between NH and CO groups (four residues up). We illustrate this by both the molecular model and the obvious helical model (**Figure 3.37** and **Figure 3.38**).

Another prevalent helical structure found is the 3_{10} helix (Figure 3.37 and Figure 3.38). As the name describes, this helix has only 3 residues per turn with a pitch of 0.60 nm. So this is a more open structure than the α-helix. It is stabilized by H-bonds between NH and CO groups (three residues up) (Figure 3.38).

Another commonly found secondary structure is the β-structure (**Figure 3.39**). As shown here, it is called the anti-parallel β-pleat structure. Note that it lends itself well to forming rather extended sheet structures. The β-structure can also form parallel sheets. These can be seen in the figure (Figure 3.39).

One sees in this illustration that H-bonds stabilize the sheet between NH and CO groups of adjacent strands (**Figure 3.40**).

Longer polypeptide chains not only form many segmental secondary structure regions, but they also rely upon other weaker forces to stabilize the overall protein structure. These forces combine to define the tertiary structure of the protein.

3.2.3.3 *Tertiary structure*

We illustrate the type of weak forces that work to define the **tertiary structure** of a protein through the diagram (**Figure 3.41**). There are basically four types of bonds.

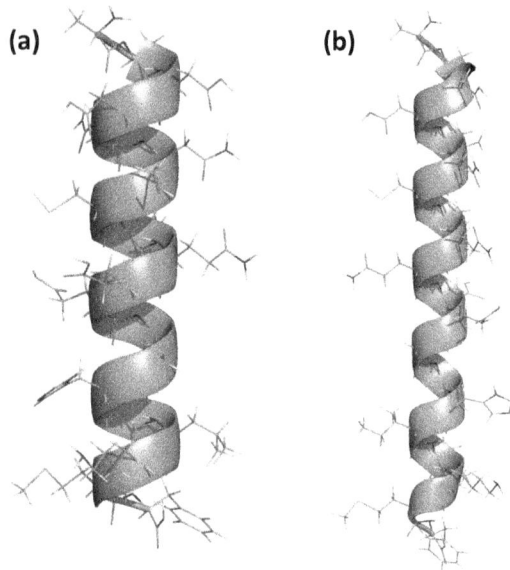

Figure 3.37. Cartoon representation of helical structures. (a) A α-helical structure. (b) A 3_{10}-helical structure.

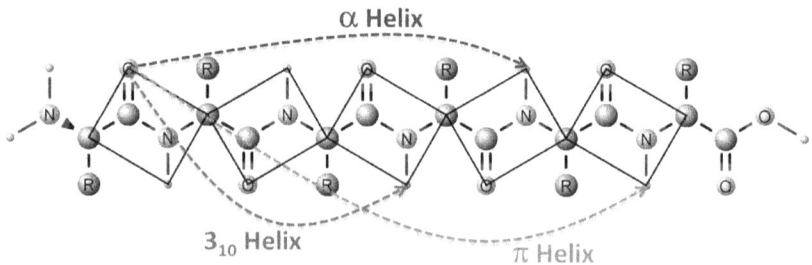

Figure 3.38. Schematic representation of the hydrogen bonding pattern in helices. The hydrogen bonding between carbonyl oxygen at position i and the amide hydrogen at position $i + 3$ stabilizes the α-helical structure (dotted lines). Similarly the hydrogen bonds between $i \rightarrow i + 2$ and $i \rightarrow i + 4$ residues (carbonyl oxygen $\rightarrow$ amide hydrogen) stabilizes the 3_{10} helix and π-helix, respectively.

β Sheets

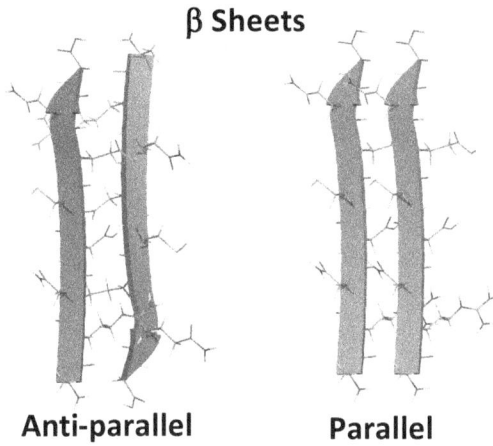

Anti-parallel   **Parallel**

Figure 3.39. Cartoon representation of β-sheets. Two β-strands can orient either anti-parallel or parallel to each other to form anti-parallel (left) or parallel (right) β-sheets, respectively. Figure generated using PyMol (DeLano, 2002).

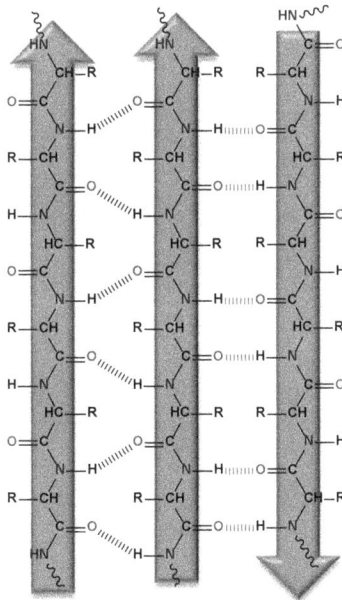

Figure 3.40. Cartoon representation of β-pleated sheets. Hydrogen bonding patterns in parallel (first two β-strands) and anti-parallel (2nd and 3rd β-strands) β-pleated sheets.

(a) The first is the *disulfide bond*. Note that sulfur exists on one of the amino acids, cysteine. Although they are not frequently in the protein chain, wherever they do exist, they seem to want to pair off for stability. This is not a particularly weak bond, but it is not very commonly encountered.

(b) Besides the many hydrogen bonds needed to stabilize the peptide chain into secondary conformations, there are also intra-chain hydrogen bonds. Intra-chain hydrogen bonds are facilitated between the structured amino acids on the surface that expose H^+ or OH^- groups. They are of strength in the 2–5 kcal/mol.

(c) Charged amino acids may find counter-ion pairing with another amino acid of the opposite charge. These are again ways to lock different amino acids into a more rigid structure by sharing the charges. The strength of these bonds is also in the 1 kcal/mol level.

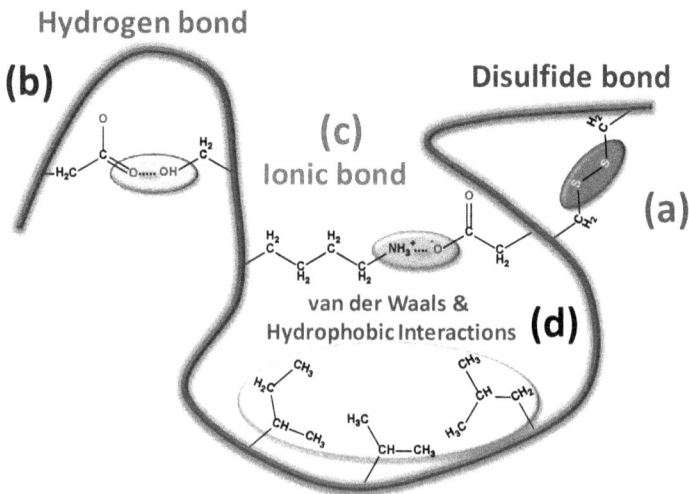

Figure 3.41.   Molecular forces stabilizing protein tertiary structure. (a) Disulfide bonds: the oxidization of the sulfhydryl groups of two cysteine residues and these are covalent bonds. (b) Hydrogen bonds: weak interactions are important in stabilizing secondary and tertiary structures (2–5 kcal/mol). (c) Ionic bonds: weak electrostatic interactions, typically side-chain involved stabilization ($\sim$3 kcal/mol) and pH sensitive. (d) Van der Waals and hydrophobic interactions: dipole molecules attract each other by Van der Waals force (transient and weak: 0.1–0.2 kcal/mol) and hydrophobic interaction, a tendency of hydrophobic groups or molecules being excluded from interacting with the hydrophilic environment.

(d) (1) The Van der Waals interaction is the weakest type of interaction. These are basically dipole–dipole interactions of very short range $(1/R^6)$. The strength of these may be 0.1 kcal/mol. (2) Hydrophobic force is a loosely defined phenomenon. This is essentially indicating that for amino acids with non-polar residues, their preference is not to be exposed to water. Thus, there is a tendency for the peptide chain to contort in such ways as to shield these nonpolar groups from the water or aqueous environment. The tendency for this has an effective energy of the weakest force, ~ 0.1 kcal/mol.

So together, these elements define the final tertiary or three-dimensional structure of a single polypeptide chain, giving it a shape and size.

3.2.3.4   *Quaternary structure*

Very frequently, even though a single unit of the protein exists in a well-defined tertiary conformation, it is non-functional until it combines with several similar proteins as an assembly. This type of collection defines the necessary **quaternary structure** that renders the protein complex functional. One of the most studied functional complexes is the hemoglobin molecule (**Figure 3.42**).

Hemoglobin is a heteromeric protein with four heme groups. Each subunit is capable of carrying an oxygen molecule. The quaternary complex renders this protein to function with allosteric (cooperative) capability, an essential requirement for the proper function of this protein.

Another unique quaternary structured protein is the filamental protein that extends within the cell. They form track-like paths on which the motor proteins move. The most prevalent of these are the *f-actin* and *microtubules*. Actin is a key component of the cellular motility network. Each actin monomer is a relatively small protein. However, the functioning protein is often with $> 10,000$ actins chained together in a two-fold helical filament (**Figure 3.43**).

Thousands of f-actin filaments align within each of the muscle cells, forming the basis of the well-ordered thin filament network within these cells. In the next section, we shall describe how filamentous myosin fibers can use these tracks to effect contractile force.

A second function of the f-actin is in the process of cytokinesis. The close association of the f-actin network on the inner surface of cell membranes provides a mechanism for cells to migrate upon the signal. As it turns out these f-actin molecules can dissociate and associate with polarity, i.e.,

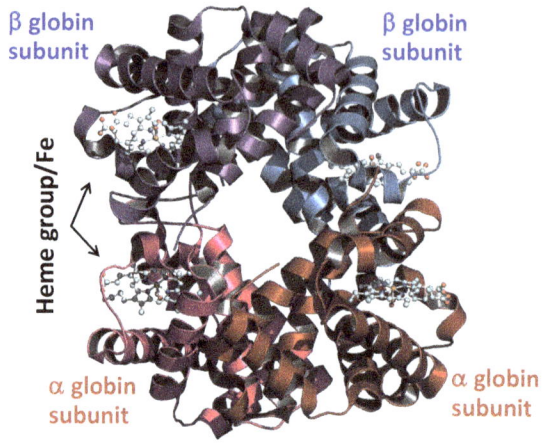

Figure 3.42. Quaternary structure of hemoglobin. Cartoon representation of hemoglobin with four subunits (α-subunits and β-subunits are shown in different shades) along with the heme group and iron are shown. Figure generated using PyMol (DeLano, 2002).

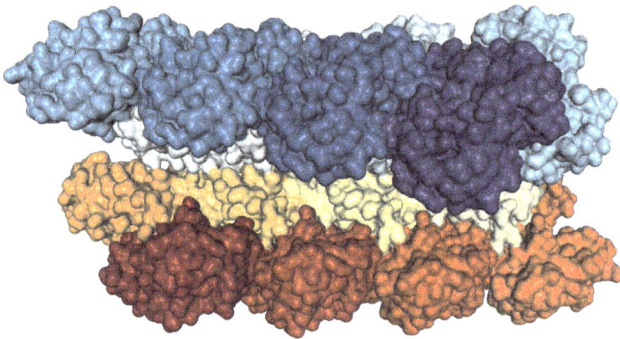

Figure 3.43. Quaternary structure of actin. Space-filled structural model of actin molecule with each subunit represented by different colors. Figure generated using PyMol (DeLano, 2002).

coming off one end of the f-actin and attaching onto the distal end. Due to the association of this f-actin to the inner membrane surface, the distortable membrane is thus guided by the association of new actin molecules.

In a similar manner, the *tubulin* monomer can associate into microtubule filaments, upon which other motor proteins, *kinesin* or *dynein* can move with polarity sense.

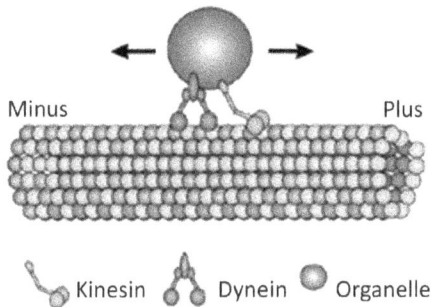

Figure 3.44. Bi-directional organelle mediated by opposing motor proteins. Dynein moves its cargo towards minus ends, whereas kinesin motors move towards the plus end. When motors are located on an individual organelle and are simultaneously active, stochastic switching between plus-end-directed and minus-end-directed motility occurs. Reproduced with permission from Steinberg (2007).

The sketch in **Figure 3.44** depicts the assembly of small tubulin molecules, A and B subunits into a filamentous microtubule. Due to the polarity difference of the subunits, the distinction will lead the filament to have a $(+)$ or a $(-)$ end, hence a polarity for the motor molecule to be guided. Kinesin and dynein are the two directed motor molecules that traverse the microtubule (Figure 3.44). Their ability to traverse these microtubules (MT) allows for intracellular transportation of molecules. We shall describe some of these systems in the next chapter. Overall, the necessary assembly of small molecules into a larger, defined complex that has a unique role is the essence of why quaternary structures are required for biological functions.

3.2.4 *Carbohydrates*

As the name sounds, carbohydrates are composed of only carbon, hydrogen, and oxygen. As such, these are very soluble substances in aqueous solutions. Simple sugars, such as glucose ($C_6H_{12}O_6$) provide the principal source of cellular energy. That is, the breakdown of the sugar is a source of cellular energy as well as components to initiate the synthesis of other cellular constituents. Simple sugars exist principally with five carbons or six carbons, called pentose and hexose respectively. The most notable of the pentose is the ribose while glucose is the best-known hexose sugar. These simple molecules can associate via the process of glycolysis, in which water is removed, and an oxygen molecule serves to link two sugars. The resulting chain of sugars is called oligosaccharides or polysaccharides depending on their length, short or long, respectively. The long-chain polysaccharides are often associated with membrane proteins, extending into the outer domain of the cell surface. Such a complex molecule is called a *glycoprotein*. These carbohydrate chains have a major role in cell signaling due to their very high affinity for hydrated molecules. The idea is that upon capture of an antigen on these carbohydrate receptor groups, the glycoprotein necessarily rearranges its conformation, which then leads to downstream signaling for requisite cellular response to the signal or insult.

This branch of carbohydrate chemistry is still evolving as we discover more and more functions for these carbohydrate groups (**Figure 3.45**).

An interesting glycoprotein complex is the antifreeze glycoprotein (AFGP), discovered by Feeney's group in circa 1970 (DeVries *et al.*, 1970; Komatsu and Feeney, 1970; DeVries *et al.*, 1971). The tight complex of this small glycoprotein presents two facets of structures: the polypeptide side and the sugar side in close proximity. The result is that such a glycoprotein can reside on interfaces. In the specific case for AFGP, the interface is the ice surface against a solution. As the namesake, this glycoprotein functions to prevent the growth of incipient ice nuclei that would otherwise deplete

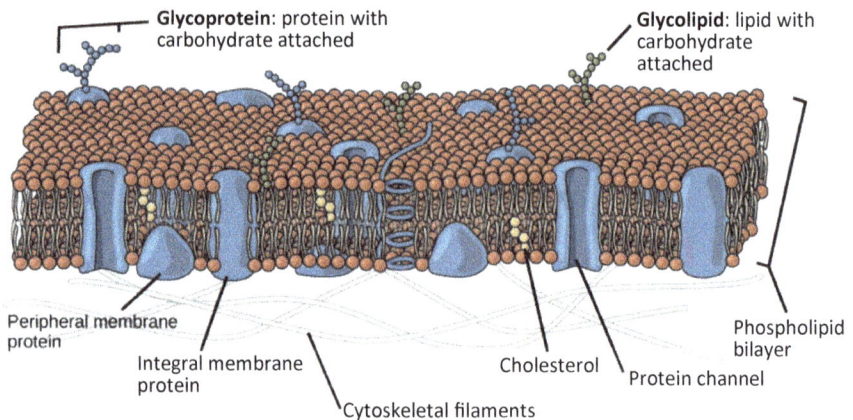

Figure 3.45.   Fluid–Mosaic Membrane Model. The fluid mosaic model of the plasma membrane describes the plasma membrane as a fluid combination of phospholipids, cholesterol, and proteins. Carbohydrates attached to lipids (glycolipids) and to proteins (glycoproteins) extend from the outward-facing surface of the membrane. Reproduced under OpenStax CNX CC BY 3.0 from OpenStax College, Components, and Structure. October 16, 2013.

the fluid component of a cell when it is submerged in sub-zero environment. A review of this topic was presented by us (Krishnan and Yeh, 2010).

We shall close this chapter and move on to some of the more specific functions of the cell that have benefitted from the use of optical methods of investigation, both as a diagnostic tool and as control machines.

For details of this section consult Cooper and Hausman (2007).

References

Atkins, P. and R. Friedman. *Molecular Quantum Mechanics*. Oxford University Press, 2011.

Cooper, G.M. and R.E. Hausman. *The Cell: A Molecular Approach*, 4th edition, Washington (DC): ASM Press and Sunderland (Massachusetts): Sinauer Associates, 2007.

DeLano, W.L. PyMOL. The PyMOL Molecular Graphics System, Version 1.2r3pre, Schrödinger LLC, 2002.

DeVries, A.L., S.K. Komatsu and R.E. Feeney. Chemical and physical properties of freezing point point depressing glycoproteins from Antarctic fishes. *J. Biol. Chem.* 245: 2901–2908, 1970.

DeVries, A.L., J. Vandenheede and R.E. Feeney. Primary structure of freezing point-depressing glycoproteins. *J. Biol. Chem.* 246(2): 305–308, 1971.

Komatsu, S.K. and R.E. Feeney. A heat labile fructosediphosphate aldolase from cold-adapted Antarctic fishes. *Biochim. Biophys. Acta* 206(2): 305–315, 1970.

Krishnan, V.V. and Y. Yeh. Structure and functional dynamics of antifreeze glycoproteins. In: S.P. Graether (ed.) *Biochemistry and Functional Characterization of Antifreeze Proteins*. Nova Science Publishers, 2010, pp. 105–140.

Staff, B. Medical gallery of Blausen Medical 2014. *WikiJournal of Medicine* 1(2): 71, 2014.

Steinberg, G. On the move: Endosomes in fungal growth and pathogenicity. *Nat. Rev. Microbiol.* 5(4): 309–316, 2007.

Tinoco, I., Jr., K. Sauer and J.C. Wang. *Physical Chemistry: Principles and Applications in Biological Sciences*. Prentice Hall, Englewood Cliffs, NJ, 1978. Zaffagnini, G. and S. Martens. Mechanisms of selective autophagy. *J. Mol. Biol.* 428(9, Part A): 1714–1724, 2016.

Chapter 4

The Expanse of Biology Problems in Need of Quantitative Solution

4.1 The Need to Measure at Increasingly Higher Spatial Resolution

As observational biology shifts toward functional biology in the 20th century, we note that the vast sets of data in the form of images of organisms as small as a cell and a virus need to be correlated with the biochemical reaction chains and cycles filling the charts of most biochemistry textbooks. The complexity of the involved processes suggests the need for approaches to dissect these complex situations into more manageable pieces. Often called the *reductionist* approach, this method gained favor over the global observational approach because it made structural sense and functional pathway significance even at the expense of not "having the whole picture." One can begin to see, however locally, how a particular function of a biochemical pathway depends on the structural interplay of specific molecules. The field of biophysics very quickly moved in to deal with the physical structures of the many varieties of biomolecules and the functional purpose each has in a strongly interwoven complex.

The traditional approach to quantitative biophysical research has been one where complex biochemical reaction pathways are studied in batch sampling modes. Typically, certain targeted molecules are labeled or otherwise identified using techniques such as nuclear magnetic resonance labels, radioactive labeling (a radioactive decay method), or fluorescence emission (see Chapter 2). Batches of assumed self-similar samples are sampled using gel electrophoresis, Western blots, and other sorting protocols to identify the labeled species at distinct times or response steps of reaction pathways. Statistical sampling leads to averaged reaction species content being measured, and hence the theories of chemical reactivity can be invoked for

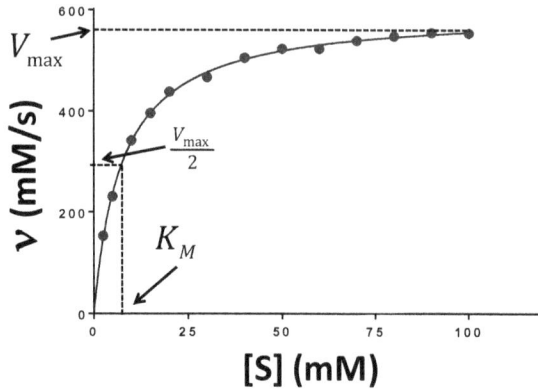

Figure 4.1.   Michaelis–Menten model. The variation in the rate of an enzyme-catalyzed reaction with concentration of the substrate. When $[S] \ll K_M$ the rate is proportional to $[S]$ and when $[S] \gg K_M$ the rate is independent of $[S]$. The significance of K_M is that when $[S] = K_M$, the rate is $\frac{1}{2} V_{\max}$ (arrows). The enzyme is effective at substrate concentrations at and above K_M.

identifying reacting species, reaction rates, and most probable pathways for reactions. In this very fruitful approach, many of the very critical biochemical pathways were deciphered. Quantitative measurements were then analyzed using the Michaelis–Menten relationship, allowing the biochemist to determine the velocity of reactions and other key reaction kinetic parameters. In its simplest form, this relationship is given as

$$v = \frac{[S]V_{\max}}{K_M + [S]} \tag{4.1}$$

where v is the initial reaction velocity, $V_{\max}$ is the maximum velocity attainable, $[S]$ is the substrate concentration, and K_M is the equilibrium constant for the reaction, as is depicted in **Figure 4.1**.

4.1.1   *X-ray diffraction and imaging*

Structural biology also made major advances in step with the biochemical kinetics approaches. The development of X-ray diffraction method to obtain data from crystallized, self-similar macromolecules in "locked" states allowed for structural views of large macromolecules. The determination of the double-helix structure of the DNA molecular chain from crystallized DNA molecules opened the entire field of genetic biology and what is commonly called the molecular biology of the gene. The ability of creating even

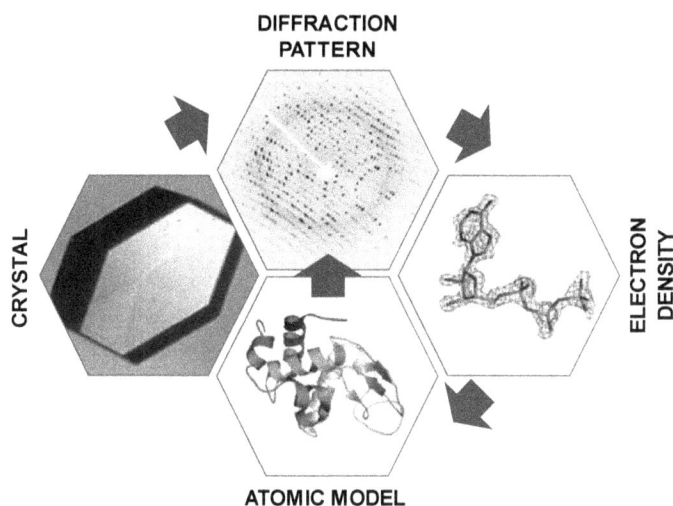

Figure 4.2. Overview of structure determination by X-ray crystallography. Diffraction patterns collected from single crystals are Fourier transformed to obtain the electron density map. Atomic models of the three-dimensional structures of the molecules are constructed from the electron density map. Final models require several cycles of refinement between the electron density map and three-dimensional structural coordinates. Figures produced under Creative Commons CC-BY License from Roedig *et al.* (2015) and three-dimensional model of 0.63 Å resolution X-ray structure of lysozyme (PDB 2vb1) using PyMol (DeLano, 2002).

micro-crystals of proteins allowed the determination of static structures of protein molecules in their various states of function. X-ray crystallography of protein structures in their various states were determined by first crystallizing these proteins and subsequently examining their diffraction pattern using X-ray diffraction apparatus. The multi-step process shows the interplay amongst the researchers of the many fields: biologist, biochemist, physicist, engineer, and applied mathematician (**Figure 4.2**).

The process is involved: the initial set of data is in the form of the diffraction patterns from these crystals. After obtaining such data from many angles and accounting for phase variances, the inverse scattering problem is carried out, leading to a first try at the electron density mapping. This electron density is then fitted to the atomic/molecular models of this particular protein/nucleic acid system. From this, feedback to the original data allows for refinement in the calculations. In such an iterative approach, many molecular crystal structures have been determined. Currently, in the Protein Data Bank (www.rcsb.org), molecular structures of many proteins have been obtained to structural resolution as low as 1–2 Å (Berman *et al.*, 2000).

As useful as this X-ray diffraction analysis method is, discernable patterns require a substantial size of the crystal of that material before the signal-to-noise becomes detectable. This becomes a problem both for the preparers of these crystals, needing to fabricate larger-sized crystals of macromolecules, and for the data analysis, needing to extract signals from a noisy background. In 1995, Sayre and Chapman (1995) already considered the possibility of conducting X-ray microscopy studies of single particles or single cells without the need for protein crystallization process. They called this idea a single-copy X-ray crystallography. Although computational signal processing methods were developed at that time, the need for intense X-ray fluence was the limiting issue. First of all, direct imaging of single molecules requires immense X-ray fluence at such wavelength because the scattering cross-section is very small, and instruments to operate at those fluence and wavelength are limited in the world. The development of Linear Coherent Light Source (LCLS) at Stanford, the DESY (Deutsches Elektronen-Synchrotron) instrument in Hamburg, and a few others around the world constitute the extent of instrument advances in that area. Researchers using these instruments have in fact succeeded in probing the structure of a few larger particles, and we shall return to these advances in a later chapter. One of the major difficulties in these studies is that the samples are being obliterated by the incident-free electron laser beam each time. Thus, in order to build up statistics to reduce errors in measurement, there needs to be some way by which replicated samples can be sent into the path of the beam in reliable and repeated fashion. This is a task that is still being advanced, and we shall return to some of the successful approaches in Chapter 6.

Another aspect of the use of X-ray methods, either crystallographic diffraction or direct imaging, for studying biological proteins is that the measurement is each in its own, a static measurement. Methods have been developed to create stationary proteins in the many states of their operational or functional pathway. Each then has to be crystallized and re-measured. Then the detailed fitting model has to be compared so as to model the specific molecular motion leading to specific functions. As we have developed the molecular constructs of a functioning cell in the previous chapter, we note that as individual molecules, each of these entities may be of the sub-microscopic scale. So how do we visualize these structural features, and even more importantly, how to monitor the dynamics of the small changes in those features? We shall come back to this issue later, to show the progress of this fascinating field of biophysical research.

4.1.2 *Electron microscopy*

Electron microscopy is another approach for shorter-wavelength probing of the material system, the cell. The latest generation cryo-EM has the electron energy and flux for investigating single molecules or viral particles. Here, samples are not crystallized into artificial protein crystals, but rapidly frozen into the surrounding aqueous medium onto an EM grid, thus in a vitreous ice environment. In a recent study, Joachim Frank's laboratory has highlighted the rapid advances in this field as they applied the cryo-EM technique to probe the intermediate structure of Ribosome with ~ 0.7 nm resolution (Fu *et al.*, 2016). In very tightly packed biological organisms, such as a viral particle (Cheng and Miyamura, 2008) the concern of ice interfering with the native environment is minimized as these viral particles are very dense and the water content is significantly less than in a normal cell. Furthermore, it is well-known that immediate water adjacent to polymers will have their hydrogen bonds distorted, hence freezing of ice is much less likely. The ability to image these particles at times as short as 140 millisecond (ms) intervals by the Frank group suggests that time resolved cryo-EM is a strong candidate as another method for determining the dynamics of macromolecular structures. The advances of this field are rapid, and we note that the 2017 Nobel Prize in Chemistry was awarded to Jacques Dubochet, Joachim Frank, and Richard Henderson.

4.1.3 *Nuclear Magnetic Resonance (NMR) spectroscopy*

Nuclei with nonzero spin possess a permanent magnetic dipole moment that precesses around the direction of the externally applied magnetic field at a rate defined by the Larmour frequency. When a sample containing magnetically active nuclei is placed into a strong magnetic field, and radio frequency (RF) radiation is applied, resonance absorption of incident radiation occurs at the frequency corresponding to the nuclear spin precession. In a quantum mechanical description, the degenerate spin states of nuclear spins are split in the presence of an external magnetic field, giving rise to absorption of radio frequency signals during the transition between the states. The absorption (resonance) frequency of the nuclei depends on the magnetic properties of the nuclei and the intensity of absorption directly proportional to the applied magnetic field. Biomolecular NMR spectroscopy focuses on the spin-$\frac{1}{2}$ nuclei such as ^{1}H, ^{15}N, ^{13}C, and ^{31}P (for DNA and RNA) because of the simplicity in the spectrum. The basis of NMR spectroscopy is to apply radio frequency pulses which perturb the population difference between the levels of the spin states $(+\frac{1}{2}$ and $-\frac{1}{2})$ and to follow the relaxation response in the form of a Free Induction Decay (FID). Fourier transformation of this time domain signal collected then provides a linear spectral response (i.e. the NMR spectrum) of the sample. A typical one-dimensional NMR spectrum identifies the chemical nature of the spin in the molecular structure in the form of a chemical shift and the number of such spin systems present in the molecule. The second most important NMR parameter is due to the indirect coupling between the nuclear spins through the electrons (hyperfine interactions), which introduces splitting in the observed chemical shifts depending on the number of coupled spins in a particular network of spins. These coupling constants are directly proportional to the magnetic properties of the nuclei.

The one-dimensional NMR spectrum obtained from a biological macromolecule contains too many overlapping peaks to permit identification of individual resonances. Multi-dimensional NMR techniques that enable the resonances to be spread out in two or more dimensions are therefore routinely used (Ernst *et al.*, 1987). Isotropic labeling of the proteins with ^{15}N/^{13}C spins to increase spectral resolution in larger proteins (> 10 kilo Dalton [kD]). Chemical shift assignments (matching the spectral signatures to each atom) are performed using a range of multi-dimensional NMR experiments. Dihedral angle restraints of the peptide backbone and side chains are derived from both chemical shifts and coupling constants. With completed sequence specific assignments, NOESY (Nuclear Overhauser Effect Spectroscopy), a two-dimensional version of the NMR experiment to map internuclear distances in proteins (Kumar *et al.*, 1980) is used to define inter-nuclear distances between the various protons. The structural constraints derived from all these experiments are then used to calculate the protein structure. During structure calculation, the chemical shift assignments and NOE estimates are refined to produce an ensemble of structures that best matches the experimental data. Restrained energy refinement of the structures in an explicitly modeled solvent is often performed to obtain the best set of final structures (Wüthrich, 1986; Billeter *et al.*, 2008). Richard Ernst received the Nobel prize in Chemistry (1991) for

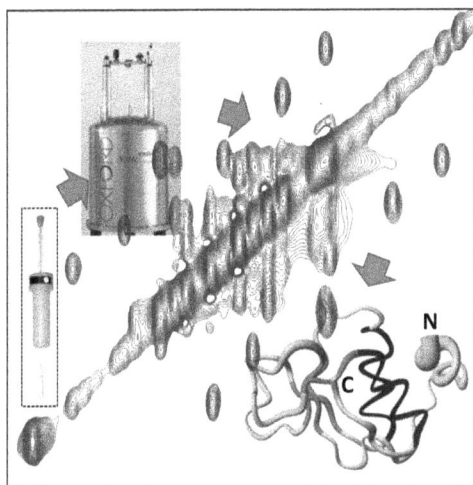

Figure 4.3. Overview of NMR structure determination. Nuclear Magnetic Resonance (NMR) spectroscopy allows determination of structures in the solution state. When the nuclear spins are polarized in the external magnetic field distinct signals from the nuclear active spins are measured. A combination of two-dimensional and three-dimensional experiments are generally collected, processed, and analyzed to obtain NMR restraint parameters that are sensitive to determine both local structural relations and events (through chemical shifts and coupling constants) as well as the global fold (via Nuclear Overhauser Effects [NOEs]) of a protein. Three-dimensional structural models generated by NMR methods also carry additional information on residue-specific dynamic motion.

his contributions towards the development of Fourier transform NMR spectroscopy and Kurt Wüthrich shared the Nobel prize for developing NMR methods for studying biological macromolecules in 2002. X-ray diffraction studies require a crystallizable protein, while NMR is suitable for macromolecules in solution. X-ray diffraction has no size limitations and provides the most precise atomic detail, while information about the dynamics of the molecule may be limited. Nuclear Magnetic Resonance excels in cases where no protein crystals can be obtained, and it provides solution-state dynamics, but in turn, delivers lower resolution structures (**Figure 4.3**).

4.1.4 *Atomic Force Microscopy (AFM) and Near-field Scanning Optical Microscopy (NSOM)*

The basis of Atomic Force Microscopy (AFM) is an approach of imaging by point scanning. Prior to the development of AFM, the point scanning technique was used in Scanning Tunneling Microscopy (STM) for material systems that have electrical conductivity. The idea is that using a very sharp ($\sim$ nm diameter) needle, one can start to map out the landscape of the surface below that needle tip. In STM, it is the tunneling current that measures the distance from the point. With an exponential decay of tunneling current, the STM became a very high resolution method for measuring the surface profiles of metallic surfaces, sometimes to a resolution of the crystalline lattice of a few tenths of a nanometer (Binnig *et al.*, 1986). Atomic Force Microscopy is an extension of the point-probe technique and has its proximal sensitivity based on the nature of the force-distance relationship. For dielectrics, where the Van der Waals force of r^{-6} dependence is the weakest force, measurements again can be achieved to nanometer resolution. With repeated crystalline structures of solids, the measured lateral resolution can be < 1 nm. Building upon this same idea, the Near-field Scanning Optical Microscope (NSOM) converts the tip of the AFM into an optical fiber. Betzig *et al.* (1986) used an optical fiber to pass light through this end tip to excite material just below the tip on the surface. Either scattering or fluorescence methods can be used to determine the structure and nature of the surface matter. There are basically two varieties of NSOM apparatus: apertured and apertureless. Near-field Scanning Optical Microscope has also been used to probe surface structures at very high resolution. Near-field optical methods exist to probe surface structures within the limits of nanometers. The scanning mode can be as fast as the AFM tip rastering, but the condition of needing the tip to be at the exact distance from the surface for accurate near-field reading of a signal has held this method back somewhat. Many feedback designs have been implemented to resolve this difficulty, but the limitation still exists, and NSOM has a lateral resolution of ~ 100 nm. Attempting to be free from the difficulty of maintaining stability in the use of near-field apertures, the aperture-free approach has been making inroads. We shall return to this topic in Chapter 5.

4.1.5 *Optical imaging*

Optical microscopy has long been the most trustworthy, the easiest to use, and the most widely used method for probing biological samples. However, as is well-known, the principle of far-field microscopy has its limitations determined by the Abbe limit. That is to say, when probing materials with a light wave of wavelength, λ, the best that one can do, using such microscopic methods, is to achieve a resolution of about $\lambda/2$. That means if one is probing with 500 nm light wave from a green laser, the best spatial resolution would be about 250 nm. We know from our previous discussions that materials that reside within a cell are all much smaller than 250 nm each. What does one do to circumvent this limitation? Within the past 50 years, three or four techniques stand out as methods to go beyond the 250 nm resolution limit.

Seeing biomolecular functional structures means to be able to identify those molecules in its functioning state. This requirement for sampling active function has added a new dimension for biophysics, and in the main, biophotonic techniques have become a new strength for probing molecular structural dynamics and function. It is in this area that progress has been made most rapidly over the past 20 years. It is no wonder that the Nobel Committee decided to award this 2014 Nobel Award in Chemistry to three active participants in this area: Eric Betzig, Stefan Hell, and William Moerner. The efforts of these individuals will be highlighted in the next chapter.

We now review a few of these biophotonic studies of functioning molecular systems, focusing on structure-dynamics interplay, using techniques perhaps just shy of the ultrahigh resolution microscopy.

4.2 Examples of How Optical Methods Are Used in Relating Structure to Function

From Chapter 3, we have outlined the structure of a single cell. Broadly speaking there are the protective membrane, the information storage in the nucleus, and the necessary workhorse in an environment that all the proteins can carry out their prescribed functions. We now focus on some essential functions. In this chapter, our emphasis is on the biological function, and the optical techniques — even those considered mature techniques — that are very often providing much of the major new insights into functional biology. We leave the new microscopies to the next chapter (Chapter 5).

4.2.1 *Protection, regulation, and specification of proper functions within the cell*

As we have seen, the most rudimentary protection of the cell lies in its membrane walls, a lipid bilayer structure that shields the entire functioning body of the cell. There must be sufficient sensory mechanisms on that surface to detect unwanted invaders, including virus and bacteria. There must be signal transduction mechanisms on that surface to regulate the flow of desired molecules, both inward and outward, as needed. The surface receptors must also be sensitive to other insults that signal appropriate downstream activities, including the need to direct locomotion and cytokinesis, to activate specific genes to initiate replication or protein fabrication, and to regulate hostile runaway conditions.

These loosely identified functions will require cooperative functioning of many proteins as well as the lipid molecules that comprise the integral membrane surface. Frequently, the signaling receptor can transmit only in a cooperative manner with other receptors signaling the same antigen binding. This process has been shown to require the dynamic interplay between the many constituents of the membrane: phospholipids (POPC or DMPC), sphingolipids (SM) and cholesterol (Chol). It is now well-established that the special constituent of *lipid raft* plays a significant role in guiding downstream signal transduction (Simons and Ikonen, 1997). We shall focus on one cellular function that is known to have its initiation signal on the cell membrane. This is the process called *cellular apoptosis*, or programmed cell death.

The idea of cellular apoptosis is that when a signal is received by that cell for its own "suicide," usually as a protection mechanism of the organism or as growth and development necessity, the cell's death must be done in a manner that the components of the cell should not be wasted. So the breakdown process should be complete, creating basic amino acids and nucleic acids, and not be wasted or scavenged by "invaders." An illustration (**Figure 4.4**) of the process of apoptosis shows that both the cell membrane and the mitochondrion play important roles for the downstream carrying out of the requisite function.

Basically the two pathways are intrinsic and extrinsic. The *intrinsic* pathway involves proteins from the Bcl-2 family and Bax family interacting on the outer surface of the mitochondrion, managing the release of Cytochrome C, which once in the cytoplasm, will interact with Caspase-9 to initiate the downstream process involving Caspase-3 as the active apoptotic initiator. The *extrinsic* pathway starts by the assembly of CD95 and CD95L (previously called Fas and FasL) on the plasma membrane. This complex affects the membrane structure, creating and releasing a secondary messenger, ceramide, downstream in the cytoplasm. Ceramide then triggers the pro-Caspase-8 to complex and create active Caspase-8. These, in turn, generate Caspase-3, to initiate the process of breaking down the cellular components.

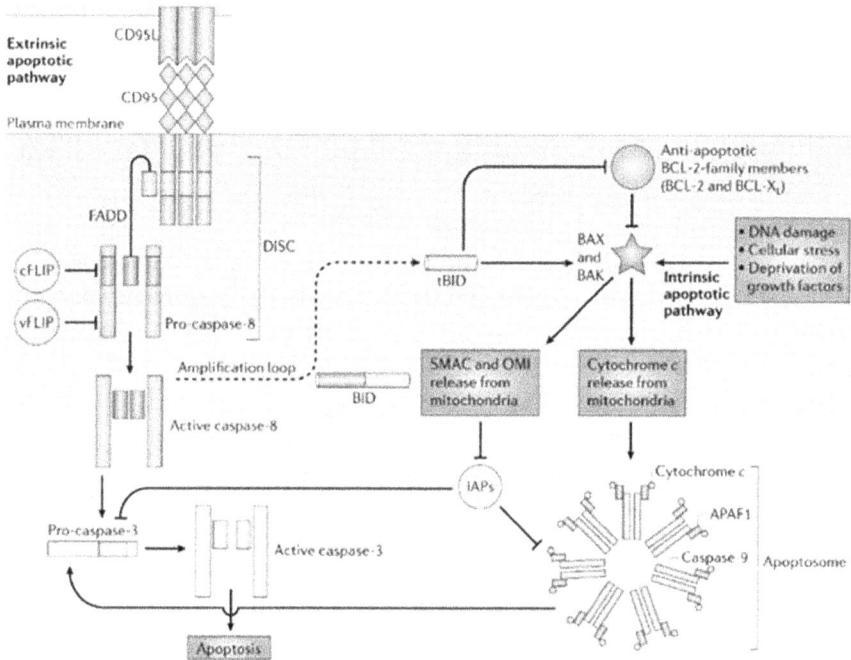

Figure 4.4. Schematic diagram of extrinsic apoptotic pathway with CD95 as an example receptor. Reproduced with permission from Siegel (2006).

One rightly asks: in all this complex of activities by small molecules, where does light enter to help in answering some important functioning questions? As in many similar situations, there is not one single optical method for the probing and diagnostics of all the steps of this biochemical process. We illustrate some experiments that have helped in answering some of the questions.

Let us take the external pathway as our example. Using fluorescently labeled cholera toxin-B (CTxB) that is known to have high affinity to lipid ganglioside GM-1, which selectively partitions into rafts thusly becoming a reliable indicator of the presence of *membrane rafts*, the cell membrane dynamics after exposure to the apoptotic agent is monitored. It is evident, by using confocal-limited spatial resolution microscopy that these GM-1 rich domains tend to coalesce upon the apoptotic trigger.

In this set of images (resolution ~ 0.5 μm), panel (a) shows a healthy retinal pigment epithelial (RPE) (**Figure 4.5**), and the micro-rafts are relatively uniformly dispersed. Upon an apoptotic insult, we see in panel (b) some domains coalescing. There are rather macroscopic domains of the GM-1 rich patches, suggesting almost platform regions. In these studies, Lincoln *et al.* (2006) further demonstrated that the domain self-assembly process is still reversible while in this initial clustering stage as evidenced by introducing methyl-β-cyclodextrin (MBCD), which depletes cholesterol and consequently leads to the disassembly of the larger raft domains.

The critical takeaway from these studies is that the lipid membrane raft seems to play an important role in process initiation. The raft domain is a membrane structure composed of POPC, SM, and Chol, at essentially equal ratios. Simons and Ikonen (1997) postulated that these groups are highly dynamic and functional, gathering or rejecting membrane proteins in order to create complexes that are functionally viable. In this case, it is the assembly of CD95 (cluster of differentiation 95) within the raft domain necessary to trigger CD95L association and the subsequent downstream apoptotic activity. Subsequent studies (Wu *et al.*, 2012) to further support this postulate was carried out by using a cellular extruder to forcibly breakup intact RPE cell membranes and create small membrane entities in a solution that contain either the platform/cluster domain composition or simply pure POPC composition. Then dynamic light scattering (DLS)

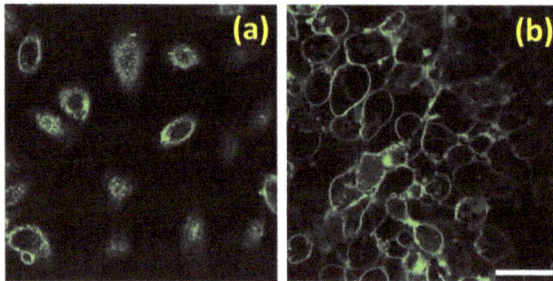

Figure 4.5. Confocal microscope images of (a) Healthy ARPE-19 (Adult Retinal Pigment Epithelial) cells. (b) UV-treated apoptotic ARPE-19 cells. Cells were labeled by BODIPY (boron-dipyrromethene) dye on C5-ganglioside GM-1. Apoptosis was induced by a dose of germicidal UV light of approximately 300 J/m^2 through a thin layer PBS buffer in the bio-hood, and then incubated with growth media for 20 hours before imaging. Bar = 40 μm. Note the finely, even punctates in healthy cells, while the apoptotic cells exhibit locations of large domains consistent with membrane raft platforms. From Wu (unpublished thesis, 2009).

was used to characterize the sizes of these extruded raft platform-containing membrane domains. A uniform size set of these vesicles were further diluted and put into an epi-illumination microscope for characterization. In order to obtain spectroscopic characterization of the distinct vesicles that contain these domains, these investigators utilized an optical trapping system to capture specific membrane vesicles, one at a time. As is well-known, the trapping intensity of the laser trap is also able to be the exciting light source of micro-Raman spectroscopy, now conducted on the specific trapped vesicle. This method is called laser trapping Raman spectroscopy (LT-RS). LT-RS was then conducted on these vesicles and compared to controlled vesicles. The results are shown in **Figure 4.6**.

Quantitative deconvolution of the observed C–C stretching mode intensities provides additional insight into average conformations of lipids enriched in the two reduced cell sub-fractions. Specifically, the ratios of C–C stretch intensities (I_{1065}/I_{1083} and I_{1116}/I_{1083}) provide a measure of relative contributions from *trans-* and *gauche-* conformers associated with the

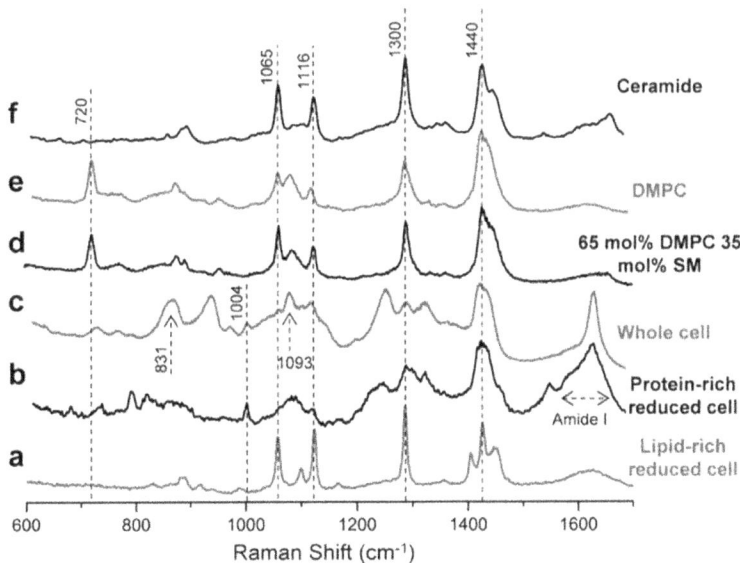

Figure 4.6. Representative laser-trapping Raman spectra of reduced cells (a, b), whole cells (c), and vesicles made from pure lipids (d–f). DMPC = 1,2-dimyristoyl-sn-glycero-3-phosphocholine, SM = sphingomyelin. Raman spectra of both classes of reduced cells are devoid of signatures associated with nucleic acid (e.g., phosphate backbone vibration at 831 cm^{-1} and 1093 cm^{-1}). From Wu *et al.* (2012).

acyl chains in reduced cells. A comparison of these ratios with those of vesicles derived from purified DMPC ($T_m = 23.9°C$), SM ($T_m \sim 39°C$)/DMPC lipids, and ceramides (Figure 4.6, d–f) at room temperature reveals that on average, methylene conformers in the reduced cells are significantly more ordered. Furthermore, it is notable that the vibrational mode due to C–N symmetric band of $+N–C(CH_3)_3$ choline head-group at 720 cm^{-1} associated with phosphatidylcholine and sphingomyelin lipids is conspicuously absent in reduced cell spectra. This observation further suggests that the reduced cell fractions not only contain more ordered lipids but they are also depleted in SM and PC-lipids. Likely candidates that substitute PC (phosphotidylcholine) lipids in these fractions include significant accumulation of cell-surface ceramides, glycosphingolipids, and gangliosides, which lack $+N–C(CH_3)_3$ moiety in reduced cell fractions.

As a further confirmation of the coexistence of raft domain and the CD95-CD95L complex, the CD95 molecules were labeled with an anti-CD95 PE (phycoerythrin) dye (red). This is then monitored with the Bodipy 488-labeled GM-1, and allowed to move in time. The time-resolved confocal images of the dynamic colocalization of CD95 and GM-1 add confirmation to the idea that the action takes place within the domains of the lipid raft. **Figure 4.7** shows the colocalization process being visualized by the regions of yellow color in this image, indicating the combination of green and red colors, to the resolution of the microscope (~ 0.5 μm).

To summarize, the mechanism of externally induced cellular apoptosis was investigated using a host of optical tools.

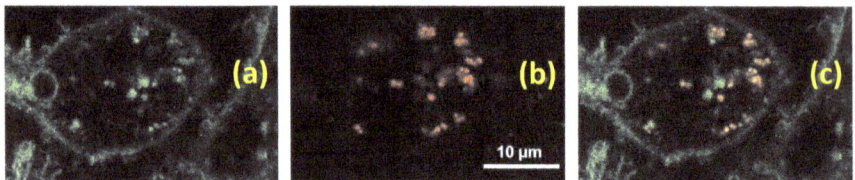

Figure 4.7.   Lipid rafts colocalize with Fas in a UV-induced apoptotic ARPE-19 cell. Image is captured by laser scanning confocal microscope (Olympus FluoView1000 MPE) with 488 nm laser excitation and 500–530 nm gated emission window for CTB-488 [(a) labeling rafts, green channel] and 620–720 nm gated emission window for Fas-PE-Cy5 antibody [(b) labeling Fas receptor, red channel]. Almost all Fas receptors (red) are colocalized with rafts domains (green), shown as mesoscopic microclusters in orange color (c). From Wu (unpublished thesis, 2009).

(1) Fluorescently labeled species, here GM-1, which has affinity to raft domains.
(2) Fluorescently labeled nuclei, which helped in the cell sorting by the fluorescent assay.
(3) Confocal microscope to monitor cell morphology.
(4) DLS to characterize vesicle sizes.
(5) Optical trapping of specific vesicles.
(6) Micro-Raman spectroscopy for structural characterization of vesicle content.
(7) Fluorescence colocalization movie showing dynamics of membrane particles.

What we see from this example is that instead of just serving as a passive medium in which certain proteins are found, elements of the lipid bilayer have been shown to phase separate into specialized platforms that foster the assembly of signaling complexes by providing a microenvironment that is specifically conducive for necessary protein–protein interactions. Villar *et al.*, (2016) states that G protein-coupled receptors (GPCRs) and relevant signaling molecules, including key enzymes such as kinases and phosphatases, trafficking proteins, and secondary messengers, preferentially partition to these highly organized cell membrane microdomains called lipid rafts. Consequently, lipid rafts are crucial for the trafficking and signaling of GPCRs. The study of GPCR biology in the context of lipid rafts involves the localization of the GPCR of interest in lipid rafts, at the basal state and upon receptor agonism, as well as the evaluation of the biological functions of the GPCR. Our above-mentioned example is but one. The continuing development of optical techniques, including FRET methods and ultrahigh resolution microscopies (to be discussed in Chapter 5) will allow for more detailed analysis of these vital processes. One asks the following questions: how do we see these molecular assemblies in action? Can optical methods provide some unique approaches for quantitative measurements of these processes?

4.2.2 *Muscle contractility*

The most obvious molecular motors within the cell are those that create motion upon being fueled by NTP. The measurement of motion is considered thermodynamic useful work. One of the most studied molecular motors is the one associated with cellular contraction. Cell division is another.

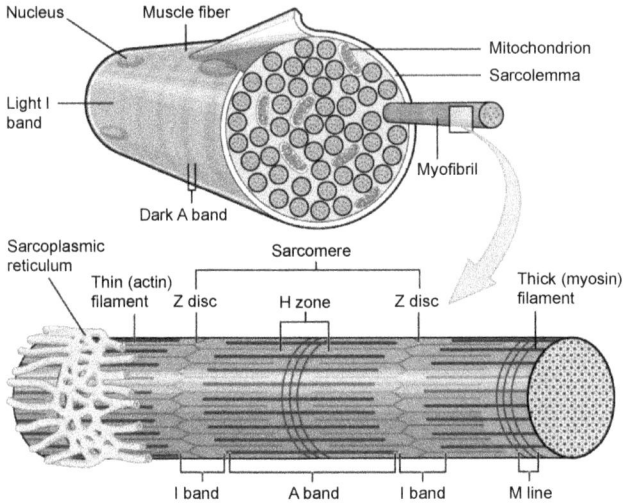

Figure 4.8. Muscle fiber. A skeletal muscle fiber is surrounded by a plasma membrane called the sarcolemma, which contains sarcoplasm, the cytoplasm of muscle cells. A muscle fiber is composed of many fibrils, which give the cell its striated appearance. Reproduced under Creative Commons Attribution 4.0 International License from https://opentextbc.ca/anatomyandphysiology.

Here, we consider cellular contraction process. In these contractile cells, the molecule receiving the excess energy will change its conformation in a way that, in association with other molecules, produce lineal translation motion, resulting in the cell's change in shape. Muscle cells and cardiac cells are the most well-known cell types. We first show a diagram of the known structure of a cross-sectional of the skeletal muscle cell (**Figure 4.8**).

It is immediately obvious that these cells are very complex with a large number of protein and membrane constituents. The contractility mechanism is composed of the highly ordered assembly of the myofibril, each of which contains lineal series of the basic unit of contractility, the *sarcomere*.

A closer schematic view of the sarcomere is shown in **Figure 4.9**. Each of the serially connected sarcomeres is composed of two types of principal contractile elements, the f-actin (thin) filament, and the thick myosin filament. The mechanism of function of the muscle cell is now fairly well established. Actin filaments have binding sites for the myosin heads to bind individually. Upon binding, the expenditure of ATP-derived energy alters the bound state conformation of the myosin head group (thus the myosin S-1 head is the motor domain of the system). This motor generates force

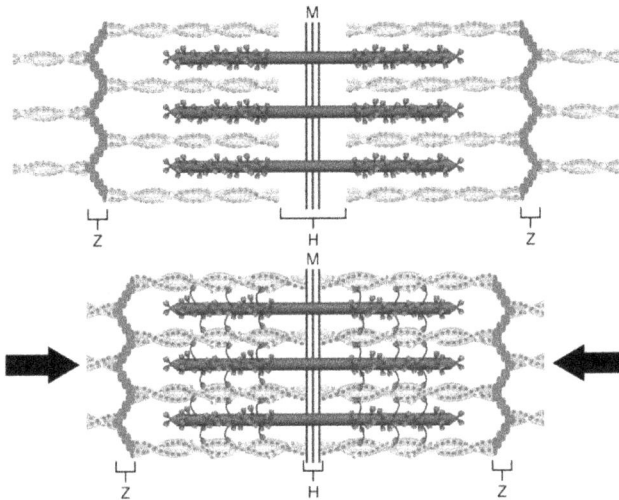

Figure 4.9. The sliding filament model of muscle contraction. When a sarcomere contracts, the Z lines move closer together, and the I-band becomes smaller. The A-band stays the same width. At full contraction, the thin and thick filaments overlap. Reproduced under Creative Commons Attribution 4.0 International License from https://opentextbc.ca/anatomyandphysiology/.

and drives on the f-actin filament to translate, resulting in a relative motion between the myosin filament and the actin filament. Because of a large number of individual molecules, each randomly delivering the driving force, the collection of all motors is a smooth macro-scale sarcomere contraction process.

The details of the contractile motor complex have been the core research focus of many investigators for at least 50 years. It suffices to say that many optical methods have played important roles in the historical progression. Note the labeling of the bands as A-band and I-band. These names were given by A.F. Huxley (Huxley and Niedergerke, 1954) because the initial microscopic observation of the sarcomere structure and its dynamics was done using polarizing interference optical microscope by this group. The globular actin monomers, when assembled, remains optically isotropic (hence I-band) while the myosin S-1 group, itself remaining mostly isotropic, is connected to highly α-helical complex of the S-2 domain and a long *light meromyosin* (LMM) tail. In the myosin thick filament, these LMM tails assemble to form high intrinsically birefringent element, thus are optically anisotropic (A-band).

Electron microscopy studies also played an early role in our current understanding of how a muscle contracts. H.E. Huxley and Hanson (Huxley and Hanson, 1954) were able to image the various states of muscle contraction using that technique. H.E. Huxley and coworkers started to use pre-synchrotron and later on, synchrotron X-ray light sources to produce diffraction from these striated structures beginning in the 1970s. We owe much of what we know about sliding cross-bridge contraction process to those investigators.

How about the details of the motor apparatus? Rayment and his group (Rayment *et al.*, 1993) first obtained the molecular structure of myosin S-1 to ~ 0.28 nm resolution using X-ray on the crystallized myosin S-1 microcrystals. This work thus confirmed many of the lower resolution studies more common to optical methods, including birefringence (Bonner and Carlson, 1975), linear dichroism (Borejdo *et al.*, 1982), diffraction order ellipsometry (Yeh *et al.*, 1987; Yeh and Baskin, 1988), FRET (Funatsu *et al.*, 1995), fluorescence lifetime studies (Burghardt *et al.*, 1983), and phosphorescence lifetime measurements (Eads *et al.*, 1984). All of these focused on the contractile apparati. The substantial evidence from the Rayment studies suggests that there exists a large-scale conformation change within the S-1 subfragment to actualize the force generation (Smith and Rayment, 1996). When these X-ray diffraction groups developed micro-crystals that captured the many states of this myosin molecular motor, including the relaxed (ATP) state, the rigor (ATP-free) state, the non-functional AMP-PNP (Adenylyl-imidodiphosphate) state, the pre-force generating state mimicked by the vanadate (intermediate state) complex, and the force producing state with ADP (adenosine diphosphate), and each was examined at the high X-ray attainable resolution, the results can be summarized in the depiction of **Figure 4.10**.

Indeed the myosin S-1 motor domain executes a conformation change upon nucleotide binding and ATP hydrolysis. In this model, one sees the location of the ATP pocket where hydrolysis must take place. More importantly, we see the binding of the actin at the distal end of this S-1. When ATP hydrolysis takes place, these studies confirm that there is a substantial conformation change for myosin S-1 tail group between the pre-power stroke state and the no-ATP rigor state. Since that distal end of S-1 is connected to the longer myosin S-2 stalk, we can project relative translation between these filaments during this process.

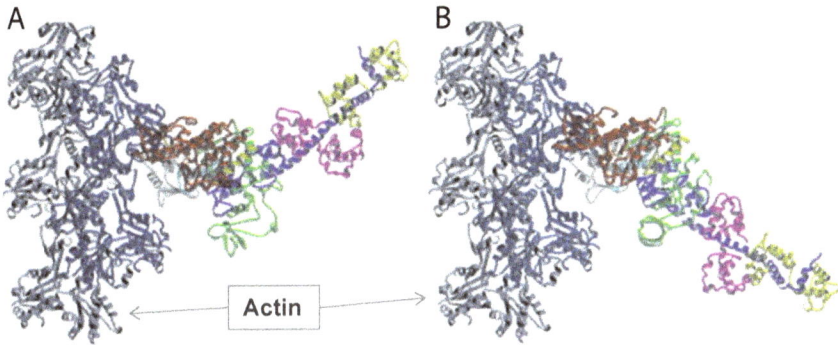

Figure 4.10. Molecular architecture of power stoke. Pre-power stroke position in (A) and the rigor position in (B), providing the extent of the swing between the extremes of the power stroke. The two shades of grey vertical structure is the actin. Reproduced with permission from Geeves and Holmes (2005).

Structural change alone does not make a function. It is necessary to show that this motion generated force is derived by ATP action on the S-1 head groups. An experiment was designed and carried out by the Yanagida group in Osaka, Japan (Ishijima *et al.*, 1998). In this experiment shown in **Figure 4.11**, many optical methods were used in conjunction.

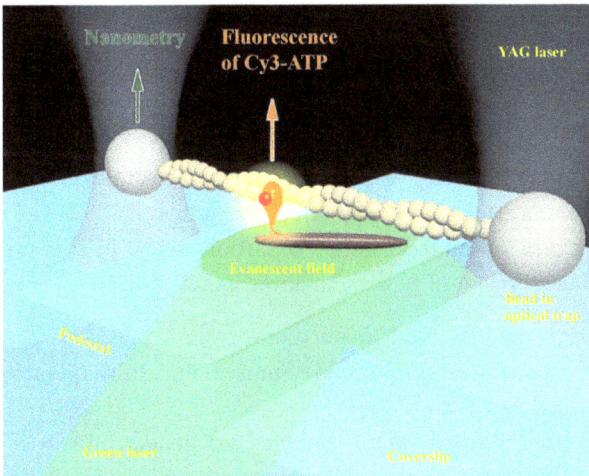

Figure 4.11. Simultaneous measurement of individual ATPase and mechanical reactions of single one-headed myosin molecules. Reprinted with permission from Ishijima *et al.* (1998).

In Yanagida's design, the goal is to measure the force produced by the myosin motor and to show that this force generation was accomplished at the time of ATP hydrolysis at the point of the single myosin S-1 molecule. The apparatus consists of a single actin filament that is biotinylated using biotin/streptavidin complex to two latex spheres in solution. The two microspheres are captured by the two-beam optical traps. One with power strong enough to overcome any force that can be generated by the molecular motor, while the other is sufficiently weak to allow the motor force to pull the bead from the center of the trap. A feedback control on the second trap then measures the amount of force necessary to bring the bead back to normal position. This construct is then positioned onto a pedestal that contains a single myosin S-1 molecule, whose S-2 subfragment is attached to the pedestal, but S-1 is free to move in solution as a tethering motion. Into the solution then is the Cy3 dye associated at the γ-position of the ATP. This dye is effectively caged, thus non-fluorescent until released by the ATP in the hydrolysis reaction. To show the presence of this dye at the location of the myosin S-1 subfragment, TIRF (Total Internal Reflection Fluorescence) microscopy is used, illuminating the S-1 domain with a laser light source that can excite the free Cy3. TIRF is useful because any active fluorophores that are further than at a distance of 100 nm away from the site of illumination will not be detected, as the evanescent wave will decay rapidly away from that surface. So the experiment calls for measurement of the simultaneous localized presence of the ADP and Cy3 fluorescence, to be synchronized with the force generation by the motor, to be measured by the weaker trap. The time trace synchronization of these events provides strong evidence of the force generation capability of myosin S-1, as a molecular motor. It is also interesting to note that pico-Newton level of forces can be measured or generated by the use of optical traps.

So let us again summarize the optical tools useful in this process of elucidating the mechanism of molecular force generation in muscle mechanics.

(1) Optical birefringence and linear dichroism.
(2) Optical diffraction of single cells of muscle and subsequent optical ellipsometry studies.
(3) Optical FRET studies of multi-labeled S-1 head groups to show dynamics of the S-1 in contraction.
(4) X-ray diffraction and analysis of the diffraction pattern at 0.15 nm, for detailed structural determination of the sarcomere.
(5) X-ray crystallography of the myosin S-1 in various states.

(6) TIRF microscopy.

(7) Optical trapping and molecular force measurement with the optical traps.

(8) Fluorescently labeled ATP as a caged compound to be released upon hydrolysis.

Where does one go after obtaining all this information about cross-bridge structure and dynamics? Clearly it focuses on the next step: how does the myosin motor take successive steps? How large are these motor-driven steps? Are there distinctions between the steps taken by different motors — different types of myosin, kinesin, dynein, etc.? What motors are involved in mitosis or cell division? These will take us into the realm of the ultrahigh-resolution microscopy in the next chapter.

4.2.3 Genetic management: Replication, search, repair, and damage control

Genetic information encoded in the DNA is stored within the cell in the duplex helical construct. In its own right, this is a complex chain, as it needs to retain the information of all of the matter that this cell needs for its total function, including transcription process for the genetic material, as well as translation processes for all proteins, membrane, and carbohydrate constructs that we have mentioned. How does this information get preserved and duplicated for next generation cells, and how do all the proteins get duplicated? These are fundamental biological questions. How do optical advances begin to play the role of elucidating and even managing some of these functions? That is the essence and purpose of biophotonics research.

As we have already seen, the preservation of the genetic material requires careful packing of the nucleic acid DNA content so that when not being called upon, they remain dormant but available. However, when called upon, they are readily available for replication or translation. In order to appreciate the complexity of such a packing and unpacking scheme, we show a schematic diagram of how eukaryotic cells pack their DNA content into the nucleus. In **Figure 4.12**, the compaction of eukaryotic chromosome is sketched in a series of levels of packing. These start with histone-induced packing, the formation of nucleosomes, the condensation into chromatin structures, and finally the formation of a chromosome.

The complex nature of the packing is in its own right, an interesting problem for structural biologists. If one next adds into this complex the need

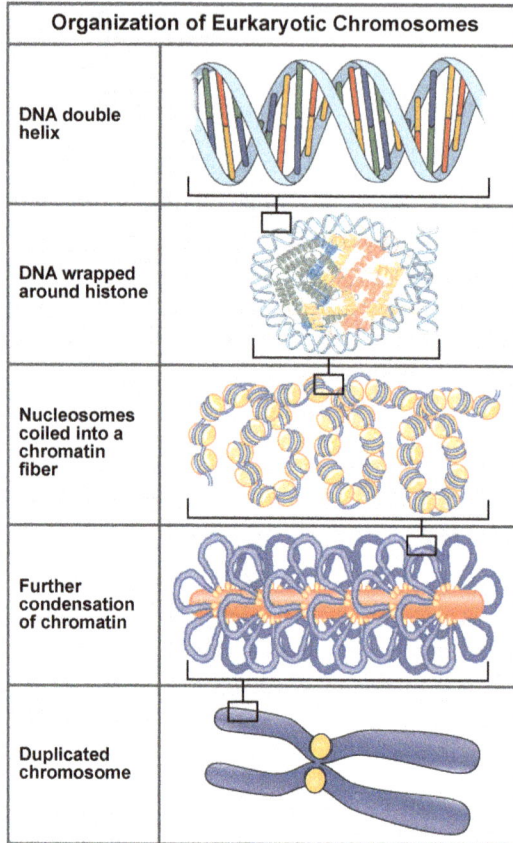

Organization of Eurkaryotic Chromosomes	
DNA double helix	
DNA wrapped around histone	
Nucleosomes coiled into a chromatin fiber	
Further condensation of chromatin	
Duplicated chromosome	

Figure 4.12.   Compaction of the eukaryotic chromosome. Reproduced under a Creative Commons Attribution 4.0 International License from openstax.org (http://philschatz.com/biology-book).

to identify a special domain within it where there is one gene that needs to be replicated, out of the presence of all others, this problem becomes a truly tangled jungle. We will limit our attempt to introduce some conceptual processes and a few successes in this set of continuing research, and in the process, we hope to understand how proteins carry out the function of gene recognition, replication, and translation.

We return to our reductionist approach to introduce the concepts and techniques. The three subsets we will introduce are helicase activity, polymerase activity, and telomere packing.

4.2.4   *Helicase activity in prokaryotic cells*

Helicase is a protein that carries out the motor function of opening up the double helix packing of the duplex DNA. This is a necessary step so that a selected domain of the DNA can become single-stranded (ssDNA) and ready for replication or translation. We choose for this example, the yeast cell that undergoes *homologous recombination* for its replication process. RecBCD is a DNA helicase that has three subunits, B, C, and D. Its role is to start with a double-stranded DNA (dsDNA) break and process it to render it ssDNA. Recall the dsDNA is held together by hydrogen bonds across strands (Chapter 3). The helicase needs to processively unwind the dsDNA to open up a domain of DNA and prepare it for subsequent combining with another homologous pair, according to the concept of homologous recombination. So helicase activity is the first step of the process. The question posed for the optical scientist is: how would you see this activity in real time?

In the laboratory of Kowalczykowski, the following set of studies was taken: it is well-known that the terminus of a dsDNA is similar to a double-stranded DNA break. Plasmid λ-DNA from *E. coli* was used. By bathing these opened λ-DNA in a solution containing RecBCD helicase, the terminal group of λ-DNA binds with this protein. The other end of this 15.4 μm long dsDNA is attached to a latex bead via usual biotin-streptavidin means.

Since the dsDNA can be labeled by intercalating dye molecules such as YOYO-1, the complex is thusly labeled to the ratio of about 1:5:YOYO-1: base pair. At this level of intercalating dye labeling, the helicase activity has not been hampered. When this system is placed in a flow system, and an optical trapping laser source is turned on, the large bead becomes trapped, and the dsDNA is elongated by the viscous drag of the flow. Since the dsDNA is labeled, it is easily visible as schematically represented in **Figure 4.13**, by an epi-optical microscope (Bianco *et al.*, 2001).

The next step is to produce the activation of the helicase by the introduction of Mg-ATP. To assure a continuous quantity of free ATP, a capillary flow cell of ~ 70 μm in its depth was used. Within this flow of the solution, the capillary walls will produce drag consistent with laminar flow conditions.

The flow was adjusted to attain truly laminar flow condition with Reynold's Number of $\sim 10^{-4}$. Under such flow conditions, even in open channel, parallel flows of different flowing contents will not mix. This approach then defined the so-called "Y-flow cells" for these experiments.

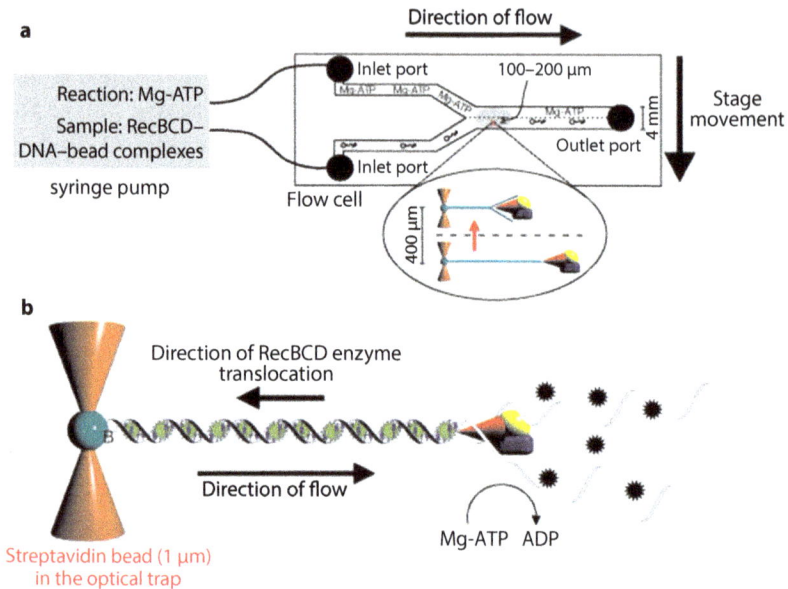

Figure 4.13.   (a) Syringe pump and flow cell: the sample syringe contains helicase–DNA–bead complexes, and the reaction syringe contains ATP. The inset points to the laser trap position, and the red arrow within the inset indicates movement of the trapped DNA–bead complex across the boundary between solutions. The inset also shows the trapped DNA with bound helicase, and its unwinding after relocation into the reaction solution. (b) Fluorescent DNA helicase assay. A trapped and stretched fluorescent DNA molecule is shown. As RecBCD enzyme translocates, it both unwinds and degrades the DNA, simultaneously displacing dye molecules (black stars). B biotinylated oligonucleotide that binds to an optically trapped microsphere. Reproduced with permission Bianco *et al.* (2001).

In the diagram of the experimental apparatus shown above, as the flow stretches the dsDNA with the optical trap holding on the bead, the Y-cell was laterally shifted from the inert channel to the active channel with Mg:ATP. The presence of the nucleotide initiates the process of helicase activity. As the helicase opens up the dsDNA, the intercalated YOYO-1 dye molecules fall into solution. Under the conditions specified, this YOYO dye is no longer fluorescent in solution, due to another mechanism of depleting the excited state of the YOYO fluorophore. So the decreasing length of labeled dsDNA is an indication of the activity of the single RecBCD helicase molecule. Thus, this optical system becomes a useful tool for probing the activity of DNA helicase under these controlled conditions.

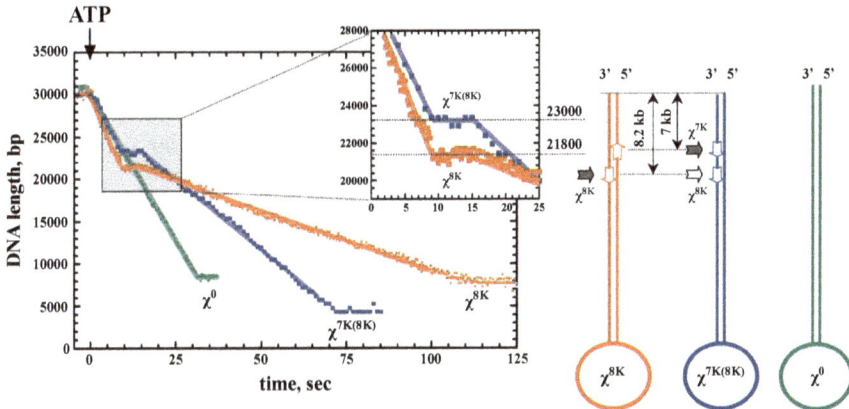

Figure 4.14. RecBCD enzyme pauses precisely at χ. Representative time courses for RecBCD-mediated unwinding of dsDNA molecules with different positions and orientations of χ. The substrates are shown schematically on the right. Black arrows point to the correctly oriented χ sequences. Reproduced with permission from Spies *et al.* (2003).

Subsequent studies by the Kowalczykowski group showed that the RecBCD molecule is a complex of two motors, B and D, with a sensor element in C (Spies *et al.*, 2003). The sensory mechanism of C recognizes the presence of special DNA domains, called the χ-domain. Sensing this domain on the $3'$–$5'$ strand allows the motors to react and adjust their speeds of translation, allowing the preservation of a single strand of the split strands. Some of the unique site recognition aspects are shown in **Figure 4.14**.

Although this scheme of optical manipulation and detection is useful for defining the process and its dynamics, it is the use of X-ray diffraction from crystallized RecBCD that showed us the high-resolution structure of this helicase molecule. The molecular structure of this molecule is given in **Figure 4.15**.

In this schematic diagram, much of the activities of the RecBCD system is shown in its molecular scale: we note the entry of the dsDNA and the sensing of the χ-domain by the RecC component first. The two single strands are then passed through the two individual channels defined by motors RecB and RecD. It is of interest to note that on one of the paths, there is a DNA nuclease, presumably chopping the ssDNA into elemental nucleotides. The other strand is preserved for the subsequent RecA binding, as part of the homologous recombination process.

In this single molecule problem, the optical techniques used include:

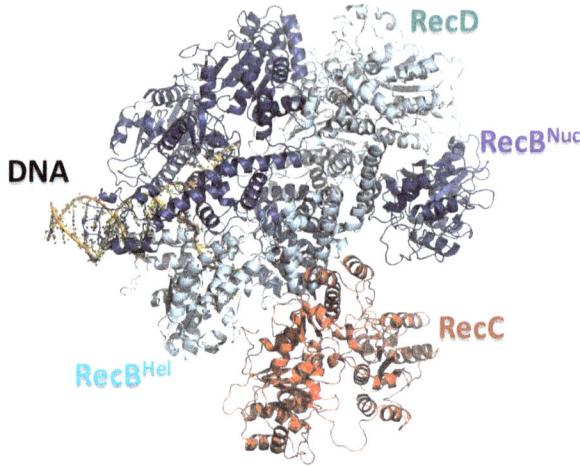

Figure 4.15. Structure of a RecBCD-DNA initiation complex. The helicase domain of RecB (RecBHel), the nuclease domain (RecBNuc), RecC, and RecD hairpin DNA substrate are shown. See details in Singleton *et al.* (2004) and structural model produced using PyMol (DeLano, 2002).

(1) Optical epi-microscopy.
(2) Optical trapping.
(3) Microfluidic device for flow and mixing control.
(4) Fluorescence monitoring as a signal of protein activity.
(5) Image quantitation for rate of activity of helicase.
(6) X-ray diffraction of crystallized RecBCD.
(7) Image reconstruction for fitting model to data.

4.2.5   *RNA polymerase-II*

The recognition that polymerase are motors that use tracks of nucleotides allows us to develop optical techniques to identify special aspects of these processes. In the RNA polymerase problem, the issue is how to examine the activity of these proteins, showing how the motor is generating force and producing a resulting strand of RNA. Schematically the process is depicted as shown in **Figure 4.16**.

The single molecule assay developed for the probing of this process by Gelles' group has many potential applications at the single molecule level. The main idea is to use an anchored polymerase. The polymerase activity is measured by the force measured at the bead that is associated with

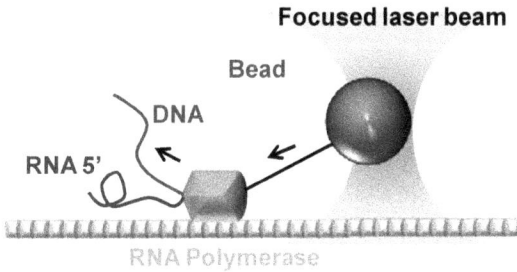

Figure 4.16.   DNA translocation by single molecules of RNA polymerase. A dsDNA is passed through the RNA polymerase that is adhered to a surface. A bead is attached to the distal end of the dsDNA. Under optical trap control, as the polymerase carries out its processive movement in reading the DNA sequence, force is generated at the bead. The necessary change in applied force to hold the bead still is a quantitative measure of the processive force of the RNA Polymerase. Adopted from Gelles and Landick (1998).

the distal end of the dsDNA, which is held by an optical trap. The motor function of the protein (the force and velocity) is measured by the force experienced by the trapped bead. As the polymerase travels up the dsDNA, it needs to locally open up the dsDNA (or helicase activity), associate a ribonucleotide to the corresponding DNA base that is being read and then move onto the next DNA base of the template strand. This moving process is a motor function which, due to the anchoring of the large polymerase, will lead to a tug on the dsDNA. This tug will be felt at the optical trap because the tug changes the bead's position from its equilibrium. The necessary increase in laser trapping intensity to recover equilibrium position is a direct measure of the force generated by the polymerase during its processive motion. Usually, these step increases in force are in the pico-Newton level. This stepwise increase of force is a diagnostic tool of the stepping activity of the polymerase II molecule (Santangelo and Artsimovitch, 2011).

In a recent report by Wang's group (Ma and Wang, 2014), a more complex DNA system is being explored using a technique that couples the tool above with additional rotation sensing movement. Because the original packing of the dsDNA comes in highly convoluted structure, it is reasoned that with the need to open up a prescribed region of the dsDNA, there will be regions beyond the point of interest that is twisted into superhelixes. Thus, the action of the polymerase is not simply force generation but also torque development. The experiment by the Wang group measured not only the force but also the angular rotation in untwisting such a superhelix (**Figure 4.17**).

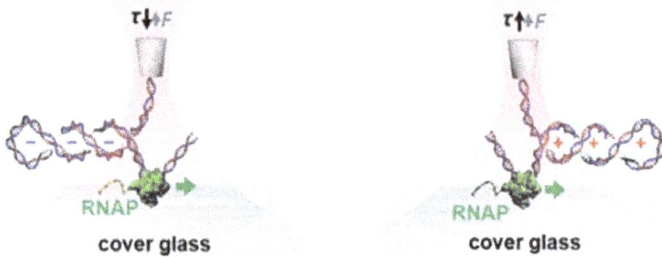

Figure 4.17.   Experimental configurations of transcription under torsion using an angular optical trap. RNA polymerase transcribes against either an upstream ($-$) DNA supercoiling (left) or downstream ($+$) DNA supercoiling (right). Artwork by Robert A. Forties, reproduced under Creative Commons Attribution-NonCommercial 3.0 Unported License from Ma *et al.* (2013).

This method then allows for single molecule torque measurement. The device is called the angular optical trap (AOT). The basis of this device is a nanofabricated quartz cylinder with the property that its extraordinary orientation is perpendicular to the cylinder axis. Oriented, this serves as the distal end element via the standard biotin-streptavidin bonding. Because quartz is birefringent, the force measurement is coupled with a change of optical ellipticity of the light signal passing through this crystal. Simultaneous monitoring leads to a direct measure of the torque of the molecule doing the twisting.

In these examples, once more we see that optical methods, in particular, optical trapping and microscopic image monitoring are useful for measurement of pico-Newton level molecular force generation.

4.2.6   *Telomere and telomerase*

The 2009 Nobel Prize for Medicine was awarded to Elizabeth Blackburn, Carol Greider, and Jack Szostak for their pioneering work in the study of *telomeres*. These are the end-caps of the DNA genetic material within the chromosome. They exist as repeated sequences of six nucleotides of DNA that carry no real genetic information. In the human telomere, the sequence is TTAGGG, hence a single-stranded sequence that is G-rich. The sequence may be repeated as many as 100 times at the end of the chromosome. Their role, as discovered by Greider and Blackburn (1985), is to become the handles for DNA polymerase as the polymerase makes duplicate copies of the useful part of the DNA for transcription. Due to the finite size of the polymerase, each replication cycle of the cell will not be able to copy the entire DNA strand faithfully. There will be a shortening of the length of the polymerase handle, which is conveniently the six-nucleotide sequence of the telomere. Thus, the "useful" nucleotides of a chromosome can withstand ∼50 to 100 replication cycles without incurring frayed "terminal DNA" that would be susceptible to damage from many types of insults. The evolutionary idea of having such a telomere also adds a check system. Indeed many organisms have a system that would signal adding on more telomere units when the sequence has become too short. A reverse transcriptase

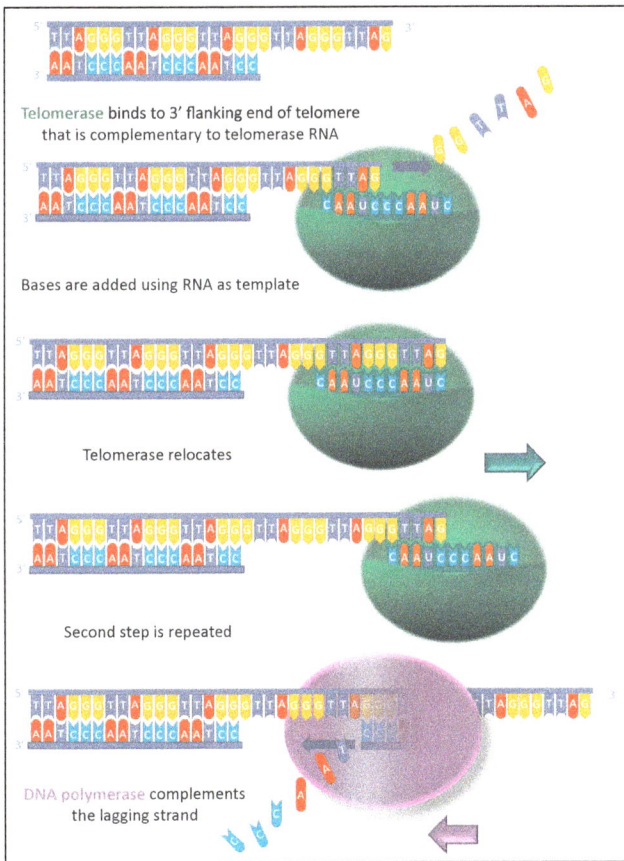

Figure 4.18.   Telomerase elongates telomere ends progressively. Reproduced under Creative Commons Attribution-Share Alike 3.0 Unported License. Attribution: Uzbas, F.

system to produce more telomeres on need is the *telomerase* (Blackburn, 2000, 2001).

This system, however unique, can be "pirated" by cancerous cells, and once the mechanism of telomerase activity goes unchecked, the cell line becomes immortal. Such cells are then cancerous! Control must, therefore, be exercised to prevent unchecked telomerase activity. It is thusly seen that the telomere management, including the controlling of telomerase activity, is an important part of healthy gene expression.

To ensure genetic homeostasis, nature has built in certain "roadblocks" against the unwinding of the terminal groups of the telomere. This is through a system of telomere binding proteins, including the repressor activation protein 1 (RAP1). The idea is that the RAP1-telomere complex will assume a conformation that renders it not able to be bound by the telomerase until a conformation change in that complex occurs. This then serves as a check and balance system. Gilson *et al.* (1994) show TEM (Transmission Electron Microscopy) images of binding regions of this protein-DNA complex. X-ray diffraction studies of crystallized RAP1 has also been useful in elucidating the structure of their DNA binding domain (Konig *et al.*, 1996). We show an AFM image of the binding of RAP1 against a 10-telomere chain. In the image shown in **Figure 4.19**, the abil-

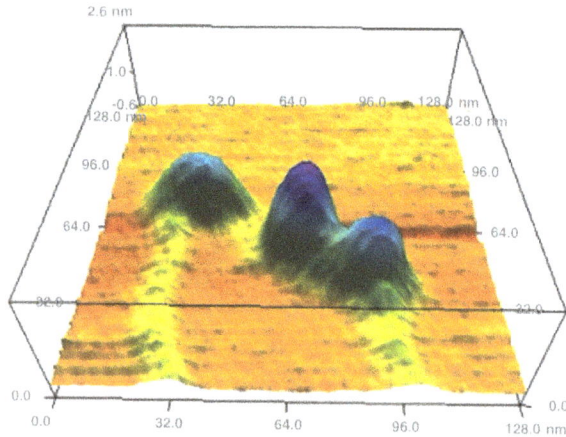

Figure 4.19. A random set of 10 telomeres in the presence of RAP1 binding protein. Protein interaction is evident in the AFM image showing peaks suggesting binding of RAP1 to telomeric DNA. Bare DNA on surface has 0.2 nm height. The peaks of the bound domains show peaks as high as 2.5 nm. Unpublished data from work of Y. Yeh, 2006 Conference presentation OWLS (2006).

ity of RAP1 to twist the telomeric DNA into a tighter complex is clearly seen. This could be vital for creating an environment that is non-conducive for telomerase binding necessary for the reverse transcriptase activity. We await newer studies to provide a quantitative measurement of the telomere capping complex.

We have outlined just a few examples of how at the single molecule level, proteins play active roles in the necessary functions of cellular survival. In each of these, many optical tools have to be brought to bear in order to even get started in our understanding of the issues of biological importance.

Some of the specific tools used amongst these studies include:

(1) Optical trapping device as a tool for manipulation and measurement.
(2) Optical polarization sensitivity.
(3) Fluorescence signal changes upon changes of the environment of the fluorophore.
(4) Magnetic trapping systems.
(5) Microfluidic chambers for environment management.
(6) Atomic force microscope for morphological checks.

Now that we have mentioned the broad spectrum of protein activities necessary to regulate, manage, and transcript DNA genes, we again ask the question, "How do we focus in and see these processes in action with sufficient spatial and temporal resolution to identify specific structure-function mechanisms?" The need for super-resolution instrumentation is definitely there. The optical scientist/engineer has to bring two other elements into the total equation for the ultimate goals. These are:

(1) The optical studies cannot be done in isolation as a "very clever or cute" experiment. The biological issues of interest are, as we have stressed in this chapter, very complex, demanding first of all, that the scientist work in a highly interdisciplinary manner, **not simply multidisciplinary**. The reason is that unless there is full understanding and appreciation by all involved of the team as to why such an experiment should be done, the piecing together of the independent data sets that are fragmented and impossible to coordinate will not advance the overall knowledge base of the problem at hand. So physical scientists must learn about the biology involved, as must the biologist learning the essence of the physical tools being employed.
(2) Optical tools are many and varied. Unless the research team is willing to take advantage of each unique feature of the totality of optical

interaction with matter, the researcher is not giving the problem its just attention. As we have pointed out in this chapter, every single biological problem takes a large number of optical tools for probing and analysis, and the need to supplement with non-optical tools is an essential requirement.

References

Berman, H.M., J. Westbrook, Z. Feng, G. Gilliland, T.N. Bhat, H. Weissig, I.N. Shindyalov and P.E. Bourne. The protein data bank. *Nucleic Acids Res.* 28: 235–242, 2000.

Betzig, E., A. Lewis, A. Harootunian, M. Isaacson and E. Kratschmer. Near field scanning optical microscopy (NSOM): Development and biophysical applications. *Biophys. J.* 49: 269–279, 1986.

Bianco, P.R., L.R. Brewer, M. Corzett, R. Balhorn, Y. Yeh, S.C. Kowalczykowski and R.J. Baskin. Processive translocation and DNA unwinding by individual RecBCD enzyme molecules. *Nature* 409: 374–378, 2001.

Billeter, M., G. Wagner and K. Wüthrich. Solution NMR structure determination of proteins revisited. *J. Biomol. NMR* 42: 155–158, 2008.

Binnig, G., C.F. Quate and C. Gerber. Atomic force microscope. *Phys. Rev. Lett.* 56: 930–933, 1986.

Blackburn, E.H. Telomere states and cell fates. *Nature* 408: 53–56, 2000.

Blackburn, E.H. Switching and signaling at the telomere. *Cell* 106: 661–673, 2001.

Bonner, R.F. and F.D. Carlson. Structural dynamics of frog muscle during isometric contraction. *J. Gen. Physiol.* 65: 555–581, 1975.

Borejdo, J., O. Assulin, T. Ando and S. Putnam. Cross-bridge orientation in skeletal muscle measured by linear dichroism of an extrinsic chromophore. *J. Mol. Biol.* 158: 391–414, 1982.

Burghardt, T.P., T. Ando and J. Borejdo. Evidence for cross-bridge order in contraction of glycerinated skeletal muscle. *Proc. Natl. Acad. Sci. U.S.A.* 80: 7515–7519, 1983.

Cheng, R.H. and T. Miyamura. *Structure-based Study of Viral Replication*, World Scientific Publishing Co. Pte. Ltd, Singapore, 2008.

Delano, W.L. PyMOL. The PyMOL Molecular Graphics System, Version 1.2r3pre, Schrödinger LLC, 2002.

Eads, T.M., D.D. Thomas and R.H. Austin. Microsecond rotational motions of eosin-labeled myosin measured by time-resolved anisotropy of absorption and phosphorescence. *J. Mol. Biol.* 179, 1984.

Ernst, R.R., G. Bodenhausen and A. Wokaun. Principles of nuclear magnetic resonance in one and two dimensions. Vol. 14, Clarendon Press, Oxford, 1987, pp. 597.

Fu, Z., S. Kaledhonkar, A. Borg, M. Sun, B. Chen, R.A. Grassucci, M. Ehrenberg and J. Frank. Key intermediates in ribosome recycling visualized by time-resolved cryoelectron microscopy. *Structure* 24: 2092–2101, 2016.

Funatsu, T., Y. Harada, M. Tokunaga, K. Saito and T. Yanagida. Imaging of single fluorescent molecules and individual ATP turnovers by single myosin molecules in aqueous solution. *Nature* 374: 555–559, 1995.

Geeves, M.A. and K.C. Holmes. The molecular mechanism of muscle contraction. *Adv. Protein Chem.* 71: 161–193, 2005.

Gelles, J. and R. Landick. RNA polymerase as a molecular motor. *Cell* 93: 13–16, 1998.

Gilson, E., T. Muller, J. Sogo, T. Laroche and S.M. Gasser. RAP1 stimulates single- to double-strand association of yeast telomeric DNA: Implications for telomere-telomere interactions. *Nucleic Acids Res.* 22: 5310–5320, 1994.

Greider, C.W. and E.H. Blackburn. Identification of a specific telomere terminal transferase activity in Tetrahymena extracts. *Cell* 43: 405–413, 1985.

Huxley, A.F. and R. Niedergerke. Structural changes in muscle during contraction. Interference microscopy of living muscle fibers. *Nature (London)* 173: 971–973, 1954.

Huxley, H. and J. Hanson. Changes in the cross-striations of muscle during contraction and stretch and their structural interpretation. *Nature* 173: 973–976, 1954.

Ishijima, A., H. Kojima, T. Funatsu, M. Tokunaga, H. Higuchi, H. Tanaka and T. Yanagida. Simultaneous observation of individual ATPase and mechanical events by a single myosin molecule during interaction with actin. *Cell* 92: 161–171, 1998.

Konig, P., R. Giraldo, L. Chapman and D. Rhodes. The crystal structure of the DNA-binding domain of yeast RAP1 in complex with telomeric DNA. *Cell* 85: 125–136, 1996.

Kumar, A., R.R. Ernst, and K. Wüthrich. A two-dimensional nuclear Overhauser enhancement (2D NOE) experiment for the elucidation of complete proton-proton cross-relaxation networks in biological macromolecules. *Biochem. Biophys. Res. Commun.* 95: 1–6, 1980.

Lincoln, J.E., M. Boling, A. Parikh, Y. Yeh, D.G. Gilchrist and L.S. Morse. Fas signaling induces apoptotic raft formation in human RPE cells that is blocked by cholesterol depletion. *Invest. Ophthalmol. Vis. Sci.* 47: 2172–2178, 2006.

Ma, J., L. Bai and M.D. Wang. Transcription under torsion. *Science* 340: 1580–1583, 2013.

Ma, J. and M.D. Wang. RNA polymerase is a powerful torsional motor. *Cell Cycle* 13: 337–338, 2014.

Rayment, I., W.R. Rypniewski, K. Schmidt-Base, R. Smith, D.R. Tomchick, M.M. Benning, D.A. Winkelmann, G. Wesenberg and H.M. Holden. Three-dimensional structure of myosin subfragment-1: A molecular motor. *Science* 261: 50–58, 1993.

Roedig, P., I. Vartiainen, R. Duman, S. Panneerselvam, N. Stübe, O. Lorbeer, M. Warmer, G. Sutton, D.I. Stuart, E. Weckert, C. David, A. Wagner and A. Meents. A micro-patterned silicon chip as sample holder for macromolecular crystallography experiments with minimal background scattering. *Sci. Rep.* 5: 10451, 2015.

Santangelo, T.J. and I. Artsimovitch. Termination and antitermination: RNA polymerase runs a stop sign. *Nat. Rev. Microbiol.* 9: 319–329, 2011.

Sayre, D. and H. Chapman. X-ray microscopy. *Acta Crystallogr., Sect. A: Found. Crystallogr.* 51: 237–252, 1995.

Siegel, R.M. Caspases at the crossroads of immune-cell life and death. *Nat. Rev. Immunol.* 6: 308–317, 2006.

Simons, K. and E. Ikonen. Functional rafts in cell membranes. *Nature* 387: 569–572, 1997.

Singleton, M.R., M.S. Dillingham, M. Gaudier, S.C. Kowalczykowski and D.B. Wigley. Crystal structure of RecBCD enzyme reveals a machine for processing DNA breaks. *Nature* 432: 187–193, 2004.

Smith, C.A. and I. Rayment. X-ray structure of the magnesium(II)·ADP·vanadate complex of the dictyostelium discoideum myosin motor domain to 1.9 A resolution. *Biochemistry* 35: 5404–5017, 1996.

Spies, M., P.R. Bianco, M.S. Dillingham, N. Handa, R.J. Baskin and S.C. Kowalczykowski. A molecular throttle: The recombination hotspot chi controls DNA translocation by the RecBCD helicase. *Cell* 114: 647–654, 2003.

Villar, V.A.M., S. Cuevas, X. Zheng and P.A. Jose. Localization and signaling of GPCRs in lipid rafts. *Methods Cell Biol.* 132: 3–23, 2016.

Wu, H. *Study of membrane dynamics with biophotonic techniques*, Thesis (Ph.D.). University of California, Davis, 2009.

Wu, H., A.E. Oliver, V.N. Ngassam, C.K. Yee, A.N. Parikh and Y. Yeh. Preparation, characterization, and surface immobilization of native vesicles obtained by mechanical extrusion of mammalian cells. *Integr. Biol. (Camb).* 4: 685–692, 2012.

Wüthrich, K. *NMR of Proteins and Nucleic Acids.* Wiley, New York, 1986.

Yeh, Y. and R.J. Baskin. Theory of optical ellipsometry from muscle fibers. *Biophys. J.* 54: 205–218, 1988.

Yeh, Y., R.J. Baskin, K. Burton and J.S. Chen. Optical ellipsometry on the diffraction order of skinned fibers. pH-induced rigor effects. *Biophys. J.* 51: 439–447, 1987.

Yeh, Y. Probing life and death struggles at the sub-cellular level. Presented at the 9th International Conference on Optics Within Life Sciences (OWLS9), Taipei, Taiwan, November 2006.

Chapter 5

Optical Microscopy

5.1 Basics of an Optical Microscope

Microscopy refers to using an instrument that can magnify the sample's structures so that one can see more details than without this instrument. A simple optical magnifying device is a convex lens. In the far field, this device can reconstruct the image of the source object, magnifying or demagnifying, depending on the curvature of the lens. In order to increase the power of magnification, complex lens designs are incorporated into the simplest optical microscope (**Figure 5.1**) (Pedrotti *et al.*, 2007). Consider, first of all, a single convex lens with focal length f_0. An object located outside of $-f_0$ and at small displacement y from the axis will have its paraxial beam focus onto this point f_0. One notes that a real image of this object is formed at the junction of that beam and another one passing through the focal point $-f_0$ and rendered paraxial at the objective lens. The total distance $f_0 + T$ is essentially the position of the real, inverted, magnified image of the object. T is defined as the *"tube length."* Magnification is given by $M = T/f_0$. At this point, we need to introduce the eye as part of the total instrument. The bare eye is a lens instrument with a photosensitive surface called the retina. The size of the eye is fairly standard, and from empirical studies, the distance of "most distinct vision" is around 25 cm. By inserting an eyepiece lens into this path, and setting the distance between the real object at $T + f_0$ to the eyepiece as f_E, the overall angular magnification compounds the objective magnification and yields a total magnification of the microscope given by

$$M = \frac{25}{f_E}\frac{T}{f_0} \tag{5.1}$$

where f_E is the focal distance of the *eyepiece*.

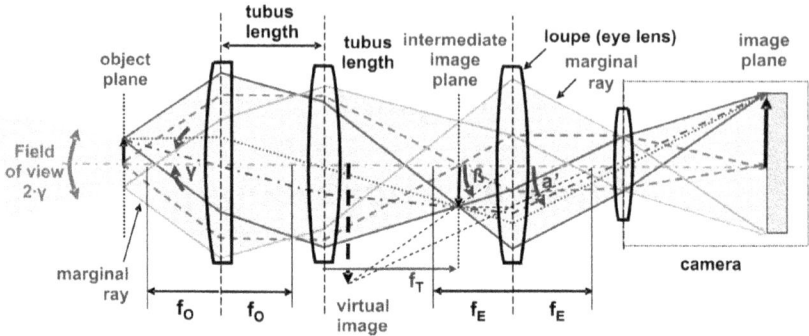

Figure 5.1. Optical path of a microscope. Reproduced under the Creative Commons Attribution-Share Alike 4.0 License.

Magnification is not the only important aspect of a microscope. Simply magnifying the image does not allow one to resolve the elements within the image. That is because each point of the object, when illuminated, will be considered a point light source producing secondary waves. These waves will then diffract and interfere. Associated with each point source will be a *point spread function* (PSF). It is the ability to resolve the nearest neighbor emitters consistent with this PSF limitation that defines the resolution of a device. Thus we have a physical optics problem, not simply a geometric optics one.

Consider light emitting from a confined circular source of radius R. When integrating over the entirety of this uniform light source, one obtains a radiance pattern that subscribes to the intensity pattern

$$I = I_0 \left[\frac{2J_1(\gamma)}{\gamma} \right]^2 \qquad (5.2)$$

where $\gamma = [(2\pi)/\lambda] R \sin \theta$. Here, λ is the wavelength of the light source; I_0 is the irradiance at $\gamma = 0$, or at $\theta = 0$. The irradiance pattern has the typical $J_1(x)/x$ distribution, where $J_1(x)$ is the first Bessel function of variable x. The first zero of this function occurs when $\gamma = 3.832$. Thus the irradiance becomes essentially zero at the point where $2R \sin \theta = 1.22\lambda$. Note that this represents the diffracted image (Airy disc) of a circular aperture of radius R. For small angles, $\sin \theta \sim \Delta \theta$, this also defines the angular radius of the Airy disc.

$$\Delta \theta_{1/2} = \frac{1.22\lambda}{2R} . \qquad (5.3)$$

One sees from this relationship that as the radius of the object gets smaller, the angular spread, or *Airy disc*, becomes larger, causing difficulty in separating small sources at the objective plane. Indeed there is a Rayleigh criterion for "just-resolvable" objects. It requires an angular separation of the centers of the image patterns to be no less than the angular radius of the Airy disc. Thus the minimum resolvable angular distance is just

$$\Delta\theta_{min} = \frac{1.22\lambda}{2R} .$$ (5.4)

Correspondingly, the minimum separation, x_{min}, of two just-resolved objects near the focal plane of a microscope with a lens diameter of D and focal length of f, is given by

$$x_{min} = f\frac{\Delta\theta_{min}}{2} = \frac{f}{2}\left[\frac{1.22\lambda}{D}\right].$$ (5.5)

The ratio D/f is called the *numerical aperture* (NA), and thus we can write

$$x_{min} = \frac{1.22\lambda}{2NA} = \frac{0.61\lambda}{NA} .$$ (5.6)

This condition shows that if an optical wavelength of $\lambda \sim 500$ nm is used, even using the highest available NA objectives (NA ~ 1.45), the minimum spatial resolution using the optical microscope is $x_{min} \sim 210$ nm. We have here the so-called Abbe limit of resolution in far-field optical resolution! This *resolution* limit for the optical microscope is the PSF, defining how much diffraction spreading will a point at focus spread. What this means is that even if one has the highest powered magnifier, a single point source on a plane will spread by diffraction in such a manner that when two point sources are within a certain distance from each other, they cannot be resolved distinctly.

In the lower panel of **Figure 5.2**, the distance between the two sources is inadequate for resolution. Mid-panel has the limit of spatial resolution (Figure 5.2).

5.1.1 *Summary of the needed parameters for imaging cellular or molecular parameters*

(1) Spatial resolution — Due to the complexity of cellular and molecular structures, the definitely higher resolution is needed to get to the detailed content of molecules as they play their role in cellular function.

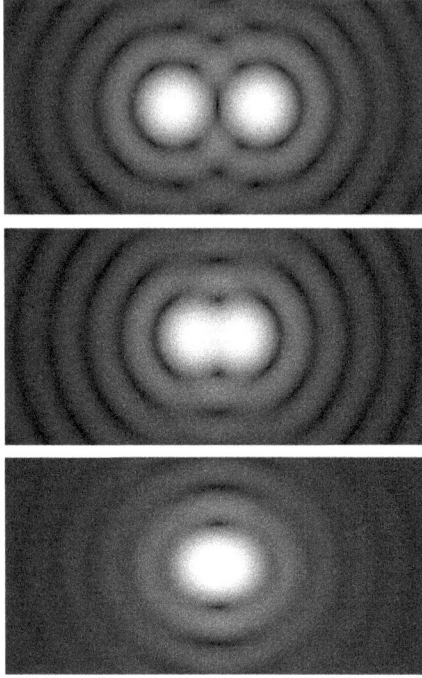

Figure 5.2. Airy disk spacing near Rayleigh criterion. Placement of two airy disks at twice the distance to the first minimum defined by Rayleigh criterion (top), at the first minimum (middle) and at half the first minimum. Reproduced under the CC BY-SA 3.0.

This is a fundamental requirement, and for visible light, Abbe limit of 200 nm has to be overcome. Molecules are simply much smaller than that dimension.

(2) Background clarity — In a conventional epi-microscope image, much of the out-of-plane, diffuse background is nonetheless projected onto the two-dimensional viewing plane (retina of the eye, a sheet of film, or camera CCD/CMOS pixel receptors). How to minimize the intrusion of those "noise" so as to extract the "signal" information is an important issue. Background rejection is a necessary must.

(3) Light traversal tracking — This is a complex issue that involves how to send light of necessary wavelength onto the point of interest, and how to retrieve the reporting light signal with the most accurate localization possible. The problem is one of light source definition. Light passage within a dense medium has a high probability of being multiply scat-

tered both in the entrance path and the retrieval trajectory from the source of interest.

(4) Enhancing the efficiency of the reporting signal — For a single photon entering into the medium of interest, and striking the reporter molecule, what is the probability of a reporter signal arriving at the detector? Often there are competing mechanisms of reporter de-excitation, minimizing the effect of the reporter from producing the signal back to the detector. The creation of highly specified fluorescent proteins via transfection protocols has allowed the tailoring of desired fluorophores at specific locations. We will discuss this topic more in the section about sources. Another aspect of managing fluorescence is to be able to control and synchronize the molecular event to the emission of the fluorescence signal. To this end, several groups have applied *optogenetic* principles to create and control the function, coordinating or synchronizing the fluorescence to the biological activity. We shall discuss this topic of optogenetics in the final chapter (Chapter 8).

(5) Time evolution of the reporter signal reflecting time sequence of molecular events — We know that molecules of interest are all dynamic. There needs to be the necessary time resolution to capture the events of interest. What methods exist to expedite the data processing for these time dependent events? This topic will be dealt with in Chapter 7.

5.2 Attempts to Meet the Needs of Higher Clarity in Microscopy

5.2.1 *Phase contrast microscopy*

For biological samples, because most of the molecules (as we have mentioned) are not with visible chromatic distinction as they do not have many intrinsically fluorescent species, the early microscope was besieged by contrast discrimination as the first order of concern. The first one to address this issue was Fitz Zernike, in the 1930s. He was awarded the Nobel Prize for Physics in 1953 for this invention.

The essence of the *phase contrast* microscope is shown in **Figure 5.3**. In the original microscope, the light source for sample illumination was delivered by a condenser lens. This allows the high intensity of the arc lamp fluence to impinge upon the sample at the location of the specimen. Discrimination of the specimen from the background is by the scattering

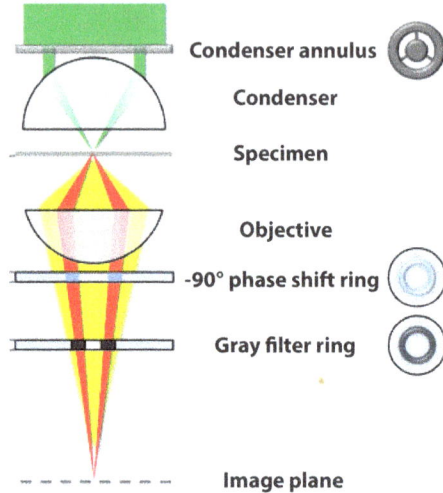

Figure 5.3. Principle of phase contrast microscopy. Illuminated light is shown in green, while the background and scattered light are shown in red and yellow respectively. Reproduced under the CC BY-SA 3.0.

of the sample versus the directly passed through the light. Since scattering efficiency of biological samples in the aqueous medium is usually weak, the image, regardless of resolution, will be of low contrast. Zernike used an annulus illumination ring instead of the broad source.

In the illustrated case, the green ring illumination alone will excite the sample specimen. The scattered light by the sample (in yellow) is collected by the objective lens, as is the directly passed through forward, unscattered illumination source. By introducing a phase ring that is out of phase with respect to the scattered light, here shown as $-90°$, although it can be others such as $+180°$, for negative or positive phase shift respectively, the image plane will exhibit a much dimmer background with respect to the scattered light. Thus the contrast of the image is enhanced. In the illustration shown in **Figure 5.4**, enhancement of contrast (right panel) is due to the application of phase contrast.

Several other approaches have been used to enhance image contrast, among them, the differential image contrast (DIC) microscope uses the polarization differential, and the dark field microscopy uses side-wise illumination. It should be noticed that these methods do not enhance the image resolution above the Abbe limit imposed by the diffraction laws.

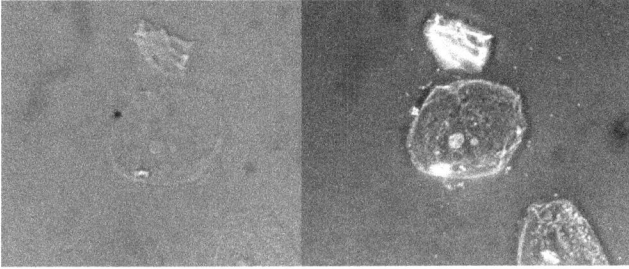

Figure 5.4. Phase contrast microscopy. Example of same cells under bright-field (left) and with phase contrast microscopy. Reproduced under the under CC BY-SA 3.0.

5.2.2 *Early X-ray microscopy imaging*

The discussion above suggests that bright field microscopy has severe limitations when it comes to imaging objects smaller than the wavelength of light. One way to overcome this limitation is to use electromagnetic radiation fields of shorter wavelength. So within the Abbe limit, using shorter wavelengths for microscopy will increase the spatial resolution, even though the techniques are the same. This gave birth to the field of *X-ray microscopy* (Da Silva *et al.*, 1992).

The reasoning behind this approach is that by using a wavelength of light that is sufficiently short, then even operating within the Abbe limit, the obtained image should provide resolution of biologically significant features. In order to achieve this task, the incident wavelength must be within a relatively non-absorbing spectral domain of the electromagnetic spectrum. Since the dominant molecule of any living organism is water, choosing a spectrally transparent water window is of paramount importance. In **Figure 5.5**, we note that a region around 4 nm is relatively free of water absorption, allowing for photons of that wavelength (energy) to access proteins and other biological compounds. These are the in the "soft X-rays" region. In the initial successful effort to image biological matter, a pulsed laser source from a large X-ray laser (NOVA) was used to illuminate a rat sperm nucleus. The scattered light was then imaged using an X-ray zone plate lens.

In **Figure 5.6**, the image of a rat sperm nucleus is seen at a resolution of nearly 55 nm. The image is enhanced by the use of antiprotamine and nano-gold labels. Here, we see the sharp boundaries of the nucleus, showing the internal wall structure of the nucleus.

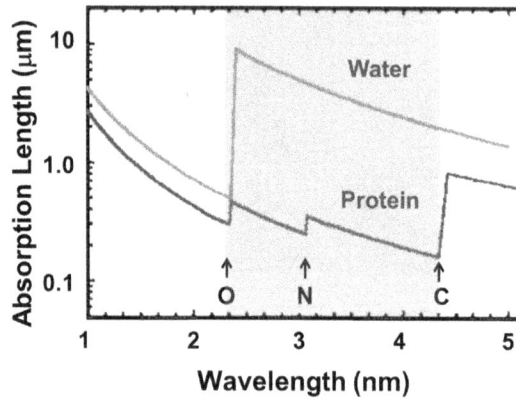

Figure 5.5. Soft X-ray window: the water window is the wavelength range between the absorption edges of oxygen ($\lambda = 2.34$ nm, $h\nu = 532$ eV) and carbon ($\lambda = 4.38$ nm, $h\nu = 283$ eV). A natural contrast within the water window is available between water and proteins (or DNA) and thus investigations on biomolecules on aqueous environment by X-ray microscopy or soft X-ray absorption spectroscopy.

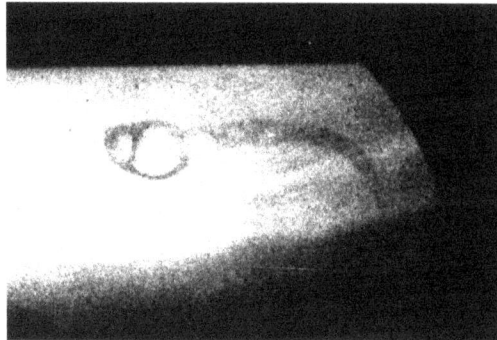

Figure 5.6. X-ray microscope image of rat sperm nuclei stained with antiprotamine and gold labeled. Details at Da Silva *et al.* (1992). Figure courtesy of Balhorn (personal communication).

The use of soft X-rays has two major difficulties for biological application: (1) it is difficult to acquire sufficient fluence for producing imaging even after the difficult task of producing lens at those wavelengths. Very few of these instruments exist in the world even now, just for the marginal needs of a microscope. (2) The X-rays are sufficiently "hard" that they damage the material samples by virtue of their relatively high-energy absorption and the creation of ionized species. Much advances have been made in this field since this original study was conducted. We will delve into the newer schemes to overcome these difficulties in Chapter 6.

5.2.3 Confocal microscopy — Beating that Abbe limit #1

The original method to overcome Abbe's limit was accomplished by Minsky in the 1950s, who developed a different version of the optical microscope that can improve the far-field optical resolution by effectively a factor of two from the Abbe limit criterion. This is the idea of the *confocal microscope.*

The confocal concept is to place a pinhole aperture at the location of the image. Rendering this pinhole only as large as the original object source point, what one encounters is effectively compounding the Airy disc effectiveness. The light intensity distribution spatially has the relationship

$$I = I_0 \left[\frac{2J_1(\gamma_1)}{\gamma_1} \right]^2 \left[\frac{2J_1(\gamma_2)}{\gamma_2} \right]. \tag{5.7}$$

Now we note that its first zero at $[J_1(\gamma)/\gamma] = 0$, and the intensity falls off at a rate faster that of the single point PSF intensity distribution. Thus, objects are now resolvable at distances closer to 150 nm, instead of 200 nm! This provides a significant improvement in the optical microscope's resolution. Beyond the resolution improvement, the confocal microscope has an added advantage, both for lateral field out of focus rejection and *depth sectioning capability* (**Figure 5.7**).

The disadvantage of this microscope is also very clear: it is a point sampling device. In order to achieve the imaging of a whole sample, the pinhole has to be scanned in the plane, one point at a time, thus rendering this device a very slow processing microscope. For 40 years since the invention of the confocal microscope, the technique was considered too slow to accomplish the needed goals of imaging cells and beyond. Several improvements have since rendered the confocal microscope one of the most desirable apparatus for any biology laboratory. The first is the ability to conduct raster scan by the use of mirror-rastering instead of having to raster the sample chamber itself, which is usually very much bulkier. Commercial laser scanning confocal microscopes are then designed with the idea of accepting incident laser light source, rastering the laser beam to create the ability to sample the entire lateral domain, and then shift the focus to essentially section the sample one layer at a time, spaced by approximately $\lambda/2$ depths.

Subsequent to this design, a multipoint sampling apparatus using a spinning disk full of the randomly spaced microlens of the Nipow design led

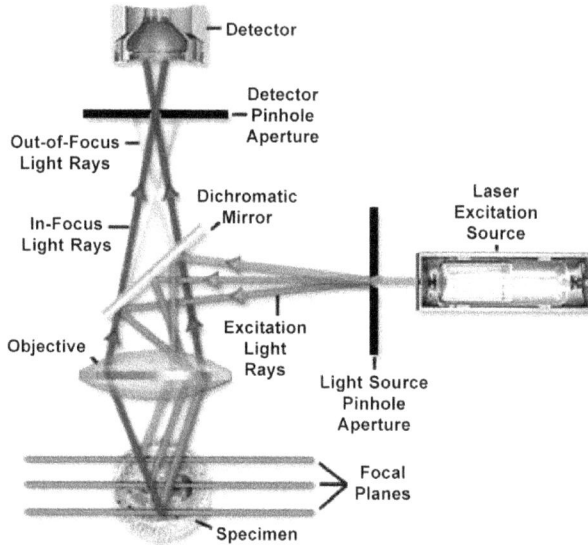

Figure 5.7. Setup of confocal laser microscope (Basil and Wassef, 2013). Reproduced under the Creative Commons Attribution 3.0 Unported License. Freely available.

to the development of the spinning disk confocal microscope. By spinning the disk, each microlens is capable of sampling one point of the object. At video rates, this basically performs the task of the raster scanning system (**Figure 5.8**), but because of the N microlens on the plane, carries out the sampling N-times faster than the single point sampling system.

We examine some of the studies made using the rapid-capture confocal scanning microscope (**Figure 5.9**) (Hubner *et al.*, 2009). In this series of images, confocal images were captured during an active biological process, here the movement of an HIV particle on the surface of a Jurkat cell. The specific manner that the infection by the HIV particle moves on the surface toward a viral synapse is traced in time. Total movement time involved here is around 80 seconds, covering a distance of 4 μm. Such time resolution suggests that temporal events of cellular dynamics can be probed by such of a microscopic method.

Given the rapid development of the high-speed computers, another method of enhancing the resolution of the confocal microscope is to perform an *optical deconvolution* of the received signals. What this method does is to take the molecular emitter's PSF distribution, and back calculate the emitter's size by performing a *deconvolution* calculation for each sampled point

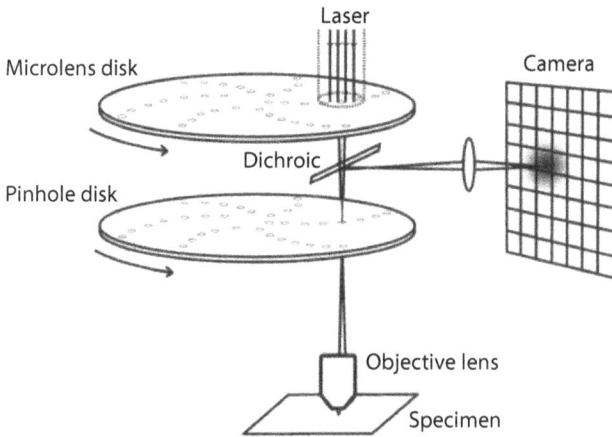

Figure 5.8. Spinning disk confocal microscopy. Aligned arrays of moving pinholes and microlenses scan a field of view in one camera exposure, giving lateral and axial resolutions typical of standard confocal microscopy. Reproduced with permission from Sisan *et al.* (2006).

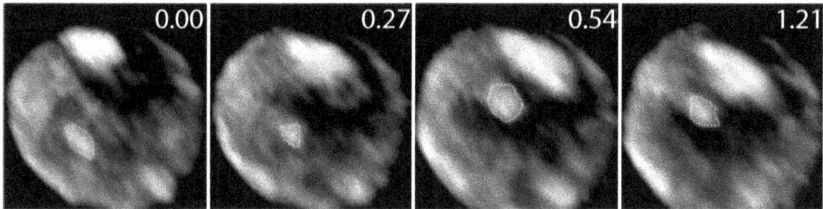

Figure 5.9. Quantitative three-dimensional video microscopy. Example of the tracking puncta of HIV Gag-iGFP at the virological synapse. Live-cell imaging was performed with a spinning disk confocal microscope using methodology. Volocity software was used to track the movement of Gag puncta moving towards a virological synapse. Reproduced with permission from Dale *et al.* (2011).

of the image. Routinely this method can reduce the resolution to ~ 100 nm, but with the significant time needed for the deconvolution algorithm to be used.

5.2.4 *Near-field microscopes — Beating Abbe limit #2*

One of the approaches to overcoming the Abbe limit is not to encounter it at all. This is the method of near-field microscopy. As we had already mentioned, the original form of near-field microscopy was the scanning tunneling microscopy, STM, invented by the IBM group in Zurich, Switzerland, resulting in Gerd Binning and Heinrich Rohrer winning the Nobel Prize in Physics for this effort in 1986. We reviewed this technique in the previous chapter.

The evolution of near-field microscopy was the development of the Atomic Force Microscopy (AFM). In this case, the surface is no longer necessarily a metal surface. The probe is a very flexible, sharp cantilever tip, usually made of SiN_3. The flexure on this tip is due to near-field forces like Van der Waals force. Thus this force has a very strong r^{-6} attractive dependence with respect to the distance of separation between surface and tip. The deflection is measured by a laser beam deflecting off the cantilever and recorded on a quadrant photodetector, where differential signals are processed.

There are two modes of AFM measurement: contact and tapping. The contact mode is effectively one where the tip is set at a small distance from the surface, and the scan is performed. Either the extent of the deflection or the force needed to counter the deflection so as to maintain a fixed distance from the surface is measured. This is then converted into a surface topology map as the tip is raster scanned (**Figure 5.10**). The tapping mode is usually used for biological samples since often the samples are too fragile to stand up to the near-contact mode. In the tapping mode, the cantilever is set into oscillation at a frequency. When the sample is being deflected differently

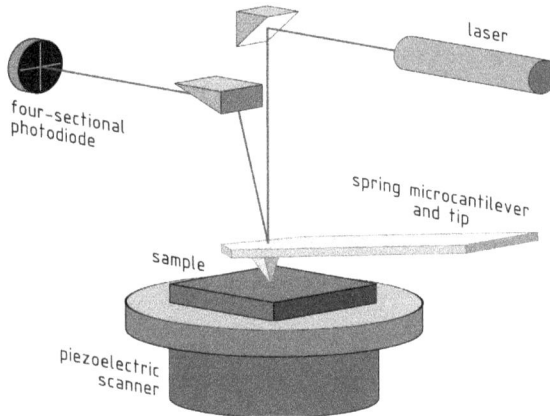

Figure 5.10.   The basic operation of AFM. A sample surface is being scanned by a tip mounted on the edge of a cantilever. The deflection of the cantilever is being monitored using a laser beam. Reproduced under Creative Wikimedia Commons Attribution.

from a free mode deflection, a change in phase or amplitude will take place. This is then recorded and converted to either topology or chemical forces. Such approaches have been successful in measuring surface morphology as low as 40 nm.

The success of the AFM spurred the push for making the AFM an optical probe. Instead of using cantilever tip, the new tips for the near-field scanning optical microscopy (NSOM) used an optical fiber shaped into a sharp tip. In many cases, the tip is coated with a silver film for better internal reflection and passage of light through this fiber in a low-loss manner (**Figure 5.11**). For detection, the light that passed from the fiber tip can be used in many ways. The work of Trautman *et al.* (1994) showed that such microscopy is able to attain a spatial resolution of ~ 100 nm. Furthermore, since the localization is provided by the input source, detection can be by a sensitive photoreceptor such as avalanche photodiode through a large NA objective for more efficient photon collection. Trautman *et al.* (1994) showed that fluorescence detection and spectroscopy could be achieved at such local regions.

Indeed, both fluorescence and scattering from the point source allows one to characterize the surface in a very local manner. The instrumentation is similar to the one used for AFM measurement.

As to be expected, the most serious concern on the use of NSOM is whether the signal was indeed coming from near-field, as it is well-known that as soon as the beam of light leaves the optical fiber, the diffraction spreading quickly renders the beam into mid-region if not far-field optical fields, making the simple analysis and certainty of near-field measurement less reliable. For that reason, many researchers in this field have made major efforts to create rapid active feedback to maintain the distance of the fiber tip from the surface.

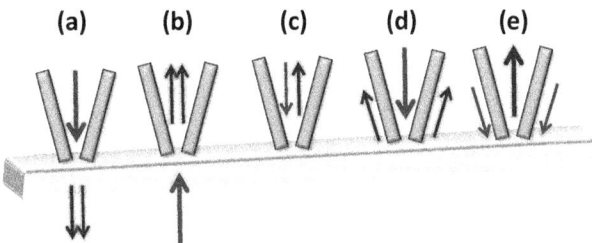

Figure 5.11.   Flexibility of aperture NSOM tips: (a) Illumination. (b) Collection. (c) Illumination and collection. (d) Reflection. (e) Reflection and collection.

A version of NSOM without aperture has also been successful. In the aperture-free design, the sample is now associated with a point while the incident light source is broadly illuminating. In order to differentiate which is the signal that is producing the scattered light, several different approaches have been used.

Recently, this process has been successfully accomplished by Tsai's group (Tsai and Lin, 2012) (**Figure 5.12**). In their scheme, a glass cover slip (GCS) has a multilayer AgOx and oxide deposition. The AgOx is within 40 nm of the surface. When illuminated by a focused laser beam (but apertureless), the AgO is photolyzed and nano-Ag particles are produced. These have free electrons that execute plasmonic oscillations, producing surface plasmons. Coupling of these via evanescent wave onto deposited particles on the other side of the GCS, those that are within evanescent wave damping layer can be excited via light tunneling. The signal derived from this would be near-field. Hence even without apertures on a tip, NSOM can be achieved. The authors caution that it is necessary to keep the incident laser power low enough so as not to permanently damage the AgOx complex.

Many of the DVD players have incorporated the NSOM features for reading high-density recorded signals on disks. The fidelity of the DVD depends strongly on this function. However, as for being able to use the technique to sample biological species on a microscope, this technique is

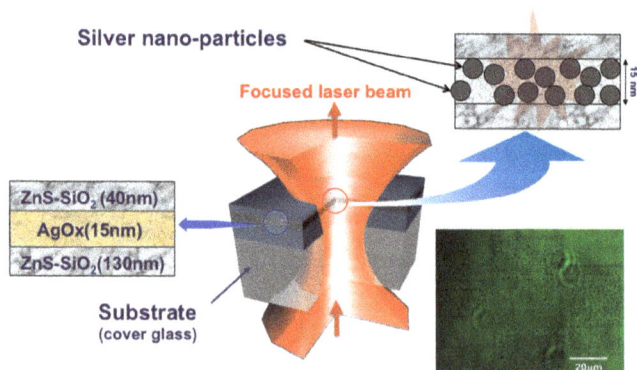

Figure 5.12.   Aperture free NSOM configuration. GCS is coated with multilayer including AgO. When excited, the photolyzed molecules become Ag nano-particles, producing surface plasmon (SP) wave. This couples in near-field mode to molecules of interest deposited on the surface of this GCS. Tunneling of light converts evanescent wave into propagating wave, producing the signal from near-field illumination. Reproduced with permission from Lin and Tsai (2012).

still in the evolving stage. The requirement that the sample stage does not move relative to the instrument is still a limiting one.

5.3   Using Contrast Agents, Intrinsic or Otherwise

5.3.1   *Fluorescence or phosphorescence emission enhances contrast*

Image enhancement above the background was achieved by the use of contrasting agents, namely fluorescent dye molecules. In early days, this process was called "staining." This paved the way for the use of fluorophores and chromophores as labels for specific sites of a molecule and increased the use of optical microscopy in biology tremendously.

Fluorescence methods rely upon the existence of a fluorophore as the basic signal-carrying entity. This occurs in two different ways. First of all, a molecule may be intrinsically a fluorophore, given the right excitation wavelength. In such a situation, the onus is to provide the proper excitation light source to excite such fluorophores. Among the amino acids, there are three that can be made to fluoresce with an excitation wavelength of 280 nm. These are Tryptophan (Trp), Tyrosine (Tyr), and Phenylalanine (Phe) (**Figure 5.13**).

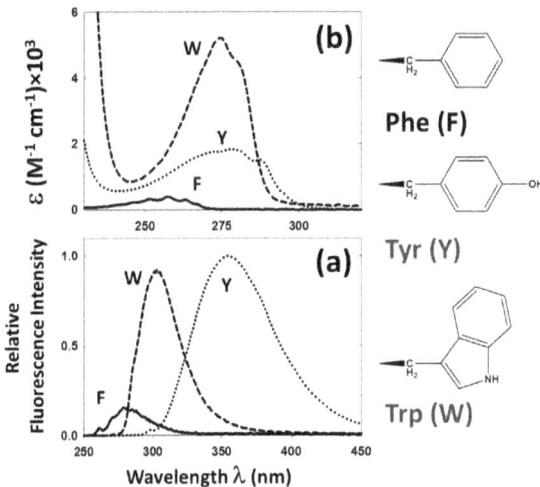

Figure 5.13.   Fluorescence profiles for aromatic amino acids, Phenylalanine (Phe, F), Tyrosine (Tyr, Y) and Tryptophan (Trp, W). (a) Excitation spectra. (b) Molar absorption coefficient.

One notes that the excitation range is from 260 nm for Phe and nearly 280 nm for Tyr and Trp. The corresponding emission peaks are 290 nm (Phe), 300 nm (Tyr), and 350 nm (Trp). Consistent with the idea we discussed in Chapter 2, the need for high efficiency in a fluorophore requires a large excited electric dipole moment. Note that all three of these amino acids possess ring structures; hence they have some floating π-bonding electrons, allowing their excited dipole moment to traverse the extent of the ring structure. Even then, the best quantum efficiency for amino acids is about 0.2, thus they are not prime candidates for probe studies.

The other difficulty in using "intrinsic" fluorophores is that each protein may have many of the same amino acid fluorophores in its structural makeup. The signal derived becomes at best the average from all of the excited fluorophores. This does not, in general, allow for highly specific characterization of the desired information related to that protein. Consequently, extrinsic fluorophores have been the main source of biological fluorescence studies.

Fluorescent dyes have long been the favorite tags for biophysical studies. The dye molecules are themselves extended ring structures with three to five rings supporting the delocalized π-electrons. Accordingly, these structures could have very large dipole moments. There are several methods to introduce fluorophores onto the molecule of interest. One way is to find a "handle" on the molecule that is not implicated for function but can be used for tagging purposes. Often terminal groups of the polypeptide chain are used to perform that task. Another frequently used handle is the unpaired cysteine bond. SH groups when not needed to bind to another SH group of the molecule is a favorite location for fluorophore tagging. Such fluorophores are *thiol-reactive*. They bind to cysteine groups. Examples are Alexa (shown in **Figure 5.14**), Texas Red (TR), Cy3, Cy5, and Atto dyes. They exhibit excellent photostability and have high quantum efficiencies (> 0.6). Similar to the protein dyes mentioned above, membranes can be labeled at their hydrophilic groups with the Bodipy dyes (**Figure 5.15**).

Note again that these have the capability of being high in quantum efficiency. An advantage of all these dyes is that they exist for visible light application, being excitable in the 400–500 nm regime and fluoresce in the 500–600 nm range.

The major disadvantage of using these extrinsic dyes for labeling is that they are structurally rather large. Adding to the fact that the handle to link them to the molecule of interest, be it membrane head group or proteins at

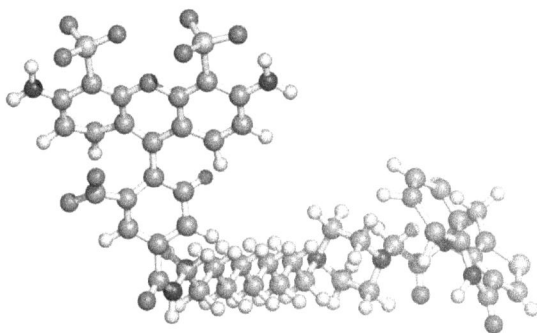

Figure 5.14. Chemical structure of Alexa-488-telenzepine (PubChem CID:91827379). Figure made using PyMol (DeLano, 2002).

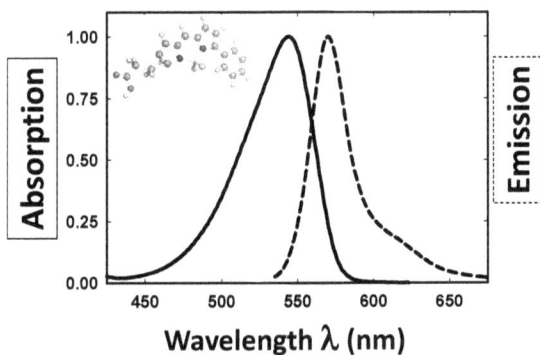

Figure 5.15. Absorption and emission spectra of Bodipy. Absorption (continuous line) and emission profile (dashed line) of Bodipy RG (ChemSpider ID 23107069) along with the chemical structure (insert) are shown. Spectral information obtained from www.spectra.arizona.edu and molecular figure is made using PyMol (DeLano, 2002).

the thiol-group, is usually a single σ-bond. That means the molecule can tumble and rotate about the bonding axis, not necessarily reflecting the motion of the tagged protein or membrane surface. Thus caution has to be exerted in the use of these labels.

Double-stranded (ds) DNA can be labeled by another method. The class of dyes listed under YOYO or TOTO are intercalating dyes that have their fluorophore groups in a plane such that they can be inserted into the planes parallel to the H-bonds that form the double strand nucleotides. In this manner, they are of extremely high efficiency when intercalated, but are effectively non-fluorescent when outside of that environment. This means

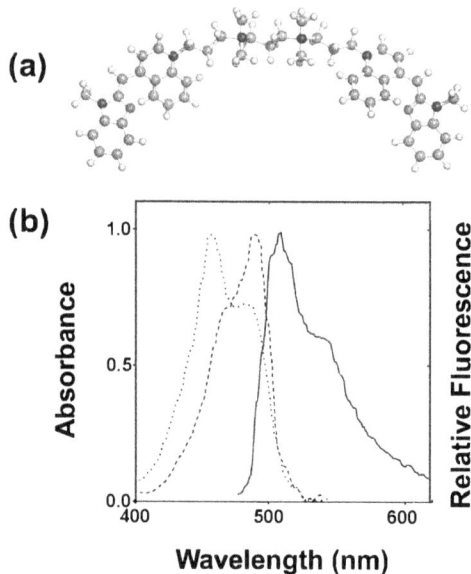

Figure 5.16. Fluorescence upon intercalation. (a) Molecular structure of YOYO-1 (ChemSpider ID 4942629) and (b) Fluorescence spectra of the YOYO-dyes. Absorption spectrum of free YOYO (dotted line), absorption spectrum of YOYO complexed with dsDNA (dashed line) and emission spectrum of YOYO complexed with dsDNA (solid line). Note the quantum yield (QE) for the free YOYO is ∼ 400 less than the DNA bound form (0.0011 versus 0.45) and therefore the emission spectrum of free YOYO is not plotted. Fluorescence spectral data were obtained from Fluorophores.org.

they are excellent for characterizing the existence of a dsDNA. An example of YOYO-1 is given in **Figure 5.16**.

In this case, the planes as shown are inserted into the dsDNA, expending some stacking energy for its stability. When the fluorophore YOYO-1 is in the ds structure, the quantum efficiency is nearly 1.0, but when in open fluid medium, $\eta < 10^{-3}$.

5.3.2   *Fluorescent proteins*

A different class of fluorophores has now taken hold in the field as the leading method of extrinsic labeling in biological samples. This is the fluorescent protein. As the name sounds, these are proteins that by themselves are fluorescent. The original native species of these proteins are from the oxoluciferase aequorin from the jellyfish *Aequorea victoria*. Once the gene of this fluorescent protein has been characterized, they are cloned and transcripted into the species that will produce a protein that terminates with the fluorescent protein. The original one absorbs at 395 nm and 475 nm and was emitting in the green (509 nm). Thus the green fluorescent protein (GFP) became the name of this labeling approach. Osamu Shimumura, Martin Chalfie, and Roger Tsien won a Nobel Prize in Chemistry for this work in 2008. The development of this type of label has become a new technology, and very quickly the spectrum has many types of fluorescent proteins, all of which can be introduced into many species that can through genetic transcription and translation, produce specified proteins with a terminal group labeled by a GFP (**Figure 5.17**).

We note that the fluorophores of the GFP variety all have a protein shell that serves as a shield structure. This renders the GFP radiation to be very insensitive to environmental change, a good thing if one wants a very reliable and robust reporting fluorophore. As mentioned in the beginning chapter, the discovery of this fluorophore has revolutionized the field of fluorescence microscopy.

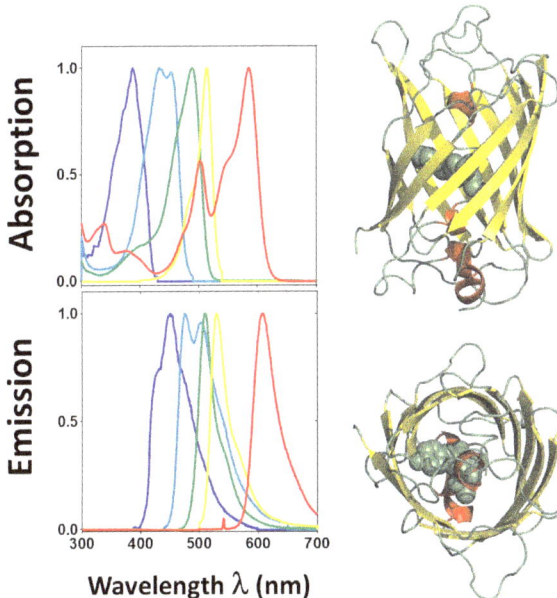

Figure 5.17. Fluorescent proteins, spectra, and structure. Absorption of emission spectra at a range of wavelengths and three-dimensional structure (pdb 1EMA). Spectral information obtained from www.spectra.arizona.edu and molecular figure is made using PyMol (DeLano, 2002).

5.3.3   *Quantum dots*

When solids with electronic band structures are excited by light, the electrons move into the upper bands and then radiate as they return to the lowest band. The radiation of this emission is a function of the energies between the lowest and higher bands. When the solid material becomes smaller in physical size, the number of atoms comprising this solid per unit cell becomes fewer, correspondingly, the bands can support fewer electrons. In the limit, when the solid approaches a single electron and nucleus, the energy structure becomes single and distinct energy levels, but still subscribing to the same macro-solid energy level spacings. In such a limit, a solid material becomes a Quantum Dot (QD). Using an appropriate selection criterion, one can select QDs that will have energies in the visible light range, thus following the established principles of the quantum theory of radiation.

The broad and high extinction coefficient coupled with sharp emissions in the visible range makes these species very efficient QD light emitters. Among the known materials that have been used to fabricate QD are Cadmium Selenide (CdSe), Lead(II) Sulfide (PbS), Lead Selenide (PbSe), Indium Arsenide (InAs) and Indium Phosphide (InP). In **Figure 5.18** we show the range of emission lines available for use.

Although the range of QD is wide, and there are major advantages in their being not easily photobleached, often these particles are toxic to

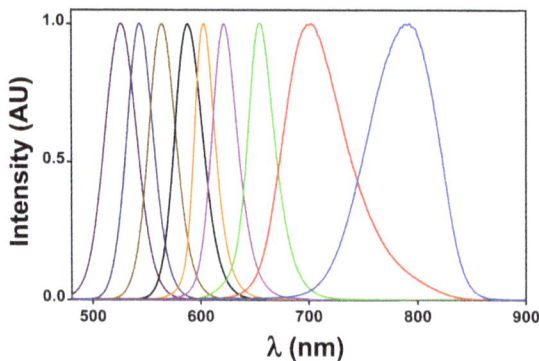

Figure 5.18.   Size dependent spectra of quantum dots. Fluorescence spectra of quanta of quantum dots with increasing size small (left) to long (right). The size-dependency scales roughly to a particle in a box model in quantum mechanics. Spectral information obtained from www.spectra.arizona.edu.

the biological system. Researchers have encapsulated these QD into self-assembled monolayers (SAM) for more non-invasive labeling. This technology is evolving as we see more and more single particle tracking experiments implemented.

5.3.4 *Search for better emitters*

From the previous discussion on quantum dots, it is clear that the ideas to push in probe development must include (1) non-toxicity to biological systems, (2) stability of the fluorescence or luminescence in the presence of many types of nearby environment, and (3) minimized photobleaching. In order to do minimal damage to the biological sample, excitation efficiency should also be very high. As it turns out, I_{sat} for rare earth emitters, such as lanthanides is very low, ~ 1.8 kW/cm^2, not like the MW/cm^2 for the dye molecules, hence more easily reached in low-dosage imaging needed for repeated excitation methods that we will discuss in the following sections. Capturing lanthanides in an inert environment is, however, a difficult feat.

Recently, exceptionally bright luminescence from silicon vacancy (SiV) at ~ 740 nm with count rates of ~ 5 million photons per second have been reported (Vlasov *et al.*, 2014) from nanodiamonds structures of size ~ 5 nm. In the laboratory of Nick Melosh, these "diamondoids," or cage-like hydrocarbon molecules where the arrangement of carbon atoms resembles the lattice of diamond are being synthesized with SiV as labels. Their size, in the range of molecular structures, and their structural stability, make these particles good prospects for the future of inert nanoprobes. Such is one aspect of the activity ongoing in the laboratory of Steve Chu.

5.4 Ultrahigh Resolution Fluorescence Microscopy

5.4.1 *Stimulated Emission Depletion (STED)*

One of the methods that succeeded using the far-field approach but attaining the ultimate resolution to being better than the confocal microscope is the *Stimulated Emission Depletion* approach, invented by Stefan Hell (Hell, 2007). The fundamental idea is that if one has already excited a fluorophore into an upper real electronic state, waiting for fluorescence emission is but one mechanism for the electrons to return to ground state. However, we can force the issue of returning to the ground state, perhaps even more rapidly than the fluorescence time of a few nanoseconds, by creating stimulated emission from that particular state.

In the scheme that Hell pioneered (**Figure 5.19**), the depletion is achieved by providing a different mode of excitation field than that which had initially populated the excited state. Assume that the initial excitation process was done using a laser in the TEM_{00q} mode; the spatial profile that was excited had a Gaussian intensity pattern. In principle, the domain of material that was excited will also be preferentially excited with the high number of excited fluorophores in the middle of the spatial pattern. Now, this spatial excitation profile has an Airy disk dimension. When a second laser is used to create stimulated emission from that excited state, the beam profile of this laser is rendered to assume the TEM_{11q} spatial structure. This is commonly known as the "doughnut" mode of the laser beam profile.

With this excitation that has been preselected to knowingly stimulate the emission of the excited fluorescent state, the shape of the spatial pattern is then rendered to overlap the original Gaussian pattern, but with the intense part of the doughnut mode saddling the wings of the Gaussian profile. In this manner, the spatially extremal part of the Gaussian excited fluorophore distribution becomes dark, leaving only the center part,

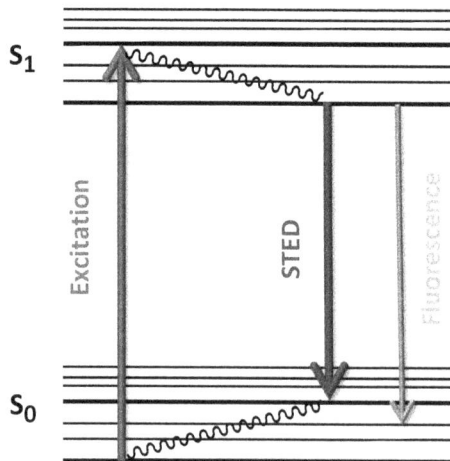

Figure 5.19. Molecules in the fluorescent state S_1 return to the ground state S_0 by spontaneous fluorescence emission. Return to S_0 might also be optically enforced through stimulated emission. To prevail over the spontaneous emission, stimulated emission depletion of the S_1 requires relatively intense light pulses with durations of a fraction of the S_1 lifetime. Adopted from Hell *et al.* (2004).

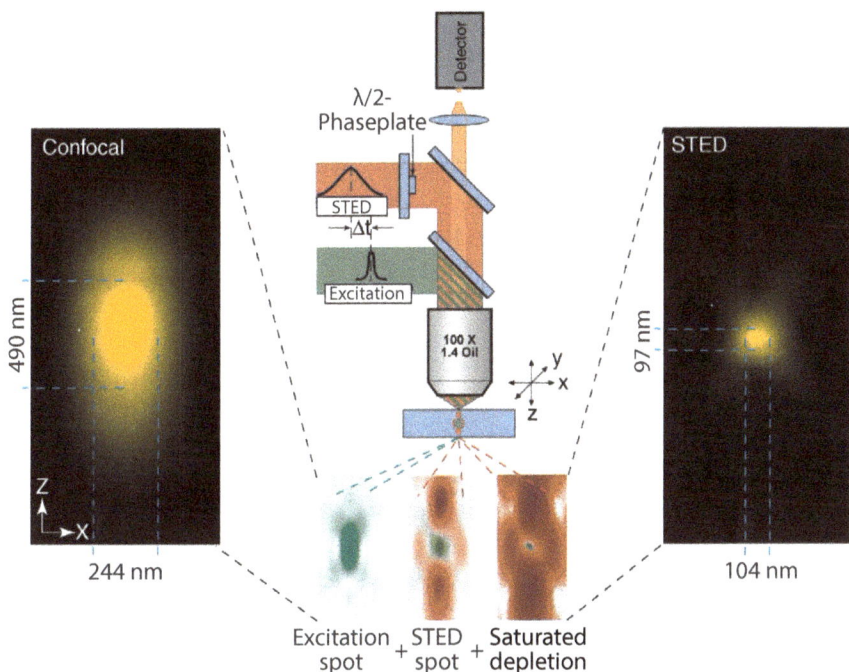

Figure 5.20. STED microscope. Excitation and STED are accomplished with synchronized laser pulses focused by a lens into the sample, sketched as green and red beams, respectively. Fluorescence is registered by a detector. Intensity distributions in the focus are shown at the bottom: the diffraction limited excitation spot is overlapped with the doughnut-shaped STED spot featuring a central zero. Saturated depletion by the STED beam confines the region of excited molecules to the zero leaving a fluorescent spot of sub-diffraction dimensions. The images on the left and right depict the confocal and the sub-diffraction-sized spot left by STED, respectively. Reproduced with permission from Hell *et al.* (2004).

which now is a "spatially clipped Gaussian" with a lateral intensity distribution controlled by the manner of the stimulating beam. In this way, the mid-domain of the excited fluorophores alone will register and spatially, they will register with a smaller spatial confine than the original Gaussian profile. In this manner, < 100 nm lateral spatial domains can be sampled (**Figure 5.20**).

$$d = \frac{\lambda}{2n \sin \alpha \sqrt{1 + I/I_s}}. \tag{5.8}$$

The concept behind this method is one of creating a spatially definable "dark state." In this scheme, a spatially emitting region of dimension

d (Eq. (5.8)) is reduced by the stimulated emission process of those fluorophores that have been stimulated to emit using a laser of wavelength λ and focused using a lens with numerical acceptance angle α in the medium of index of refraction n. As intensity, I increases above I_s, where I_s is defined as the level of light excitation that will reduce the molecule from an emission state to a non-emitting "dark" state, d is diminished. The idea is to find systems that will need a low level of I_s, so that the system is not excessively excited. Note that if the domain of excitation and fluorescence can be switched from "on" to "dark" at a rapid rate, such a method will allow for time resolved events to be followed. We shall return to this topic in Chapter 7.

5.4.2   *Photo-Activated Localization Microscopy (PALM) and STochastic Optical Reconstruction Microscopy (STORM)*

The ability to shape the excitation beam profile as developed by Hell stimulated other researchers in the field to create other schemes that would allow methods to simply "pick off" the center of the beam structure. A different approach to avail a "centroid" pick-off is based on the random blinking of fluorophores. In a study reported by Moerner (1997), the random blinking of fluorophores (on and off) was put to use for single molecule studies. If one reduces the number of the excited system to be very few, one can examine them in their individual randomness. This was true in spectrally broadened line that was inhomogeneously broadened. In the spatial domain of the fluorophore excitation, the same idea is applied. If one randomly excites the set of fluorophores and lets it recover rapidly, one can excite another round and build up statistics at specific locations of those fluorophores. In doing so, the centroid is able to be located with more certainty than the overall PSF distribution.

In the method developed by E. Betzig (Betzig *et al.*, 2006), the idea is to create a multitude of independent excitation within the same domain. This way, since each excitation is an independent event, the probability of excitation at that specific location will have a standard Gaussian distribution profile. If the number of times this fluorophore were excited is increased, the Gaussian statistical distribution of this fluorophore tightens. Therefore sampling a large number of times independently leads to a better *centroid* definition, consistent with the idea that the standard deviation of a random Gaussian variable will decrease if the number of independent events increased significantly.

The mathematical description of the idea is couched in this equation:

$$\sigma^2 = \frac{r_0^2 + q^2/12}{N} + \frac{8\pi r_0^4 b^2}{q^2 N^2}. \tag{5.9}$$

Here, σ^2 is the actual standard deviation resulting from the number N photons collected, while r_0 is the standard deviation of the point spread function (PSF), $q =$ the size of an image pixel, and $b =$ background noise per pixel.

To experimentally realize this feature, Betzig developed a way to excite a set of fluorophores in a domain with stochastic repetition, but reliably be certain that it is that domain that has been excited. In *Photo-Activated Localization Microscopy (PALM)*, the approach is to create a secondary channel for de-excitation (into the "dark" state) of the excited fluorophore but in this case, temporally and very rapidly. To achieve this end, the sample molecule is sparsely labeled with a particular fluorophore. Excitation with a weak pulse of light renders a partial set of these fluorophores fluorescent, while the others remain in the dark state. Sequential pulses can stochastically excite other dark-state fluorophores and make them emit, while the others enter dark state. This continues until all the fluorophores are photobleached. In the f-PALM configuration, two lasers are used to create the image. One is the excitation or activation pulse; this creates excitation of the fluorophores that will subsequently photobleach and disappear from the scene. The second, a weaker excitation source is used to be the constant monitor of the events. Random excitation and photobleaching, followed by new fluorophores being subject to the same repetitive process leads to a large number N. This is statistically sampled and tabulated for spatial localization.

The idea is schematically illustrated in **Figure 5.21**. In this particular experiment (Shroff *et al.*, 2007), two different fluorophores have labeled a specimen. Selected weak excitation of each set creates PSF localization of many of the spots, some of which are not resolved. But repeated de-excitation and excitation for extended periods of time allowed the PSF localization to be realized. Image reconstruction protocols can then be used to form the original specimen based on these localized points.

A variation to the PALM scheme called STORM was developed by Xiaowei Zhuang and associates (Rust *et al.*, 2006). In this version, STochastic Optical Reconstruction Microscopy, a pair of fluorophores that can trig-

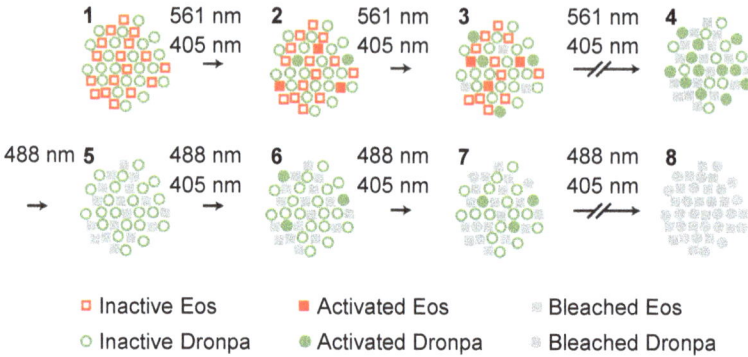

Inactive Eos **Activated Eos** **Bleached Eos**
Inactive Dronpa **Activated Dronpa** **Bleached Dronpa**

Figure 5.21. A specimen initially expressing inactive EosFP and Dronpa molecules (step 1) is exposed to a 405-nm activation light and a 561-nm Eos-excitation light (steps 2 and 3) until all EosFP molecules are detected, localized, and bleached (step 4). The many active Dronpa molecules that then exist (step 4) are deactivated by using an intense 488-nm light (step 5). Both 405-nm and 488-nm light are then applied (steps 6 and 7) to serially activate, detect, localize, and eventually bleach (step 8) all remaining Dronpa molecules. PALM images encompassing 105 to 106 molecules are thereby acquired, typically in 5–30 minutes each. Reproduced with permission from Shroff *et al.* (2007).

ger the other into its dark state is used, randomly arranged in the macromolecule of interest. Stimulating one set fluorophore at random elicits fluorescence response from the other randomly spaced set. In their case, it is the Cy3-Cy5 pair. After many repeats, the statistics build up satisfies the previous equation (Eq. (5.9)) as far as localization of the centroid of the emitters. In this manner, the spatial domain of the labeled large molecule of interest is ascertained.

The utilization of a distinctive pathway to bleach and recover rapidly is used in another version of STORM. In this method, the single fluorophore is in an environment that can spontaneously be switched to the dark state. Here, the chromophore, upon single laser excitation, immediately blinks off via a triplet state electron energy transfer. The presence of oxygen in the intracellular environment brings the triplet state system back to ground state, thus preparing it for another cycle. Such an approach is named *direct*-STORM or d-STORM. After many cycles, data reproduce the labeled region. Several hundred fluorescence photons are detected per individual fluorophore before they are photobleached or photoswitched. An excellent review of these methods can be found in Sauer (2013).

The advantages of these two variants over that of the STED microscope is that here, the excitation is done in the wide-field. Thus, a single excitation may cover many independently-located fluorophores. The repetition just tightens up the standard deviation, and thus produces a high degree of spatial localization. Overall, the idea is the same as the STED one, which is active control of the region of those excited fluorophores. This process is based on sampling by stochastic excitation/de-excitation.

The process of building up a set of data is often very slow. Many biological processes have time dependence that precludes sampling using these methods. So there is the need for something that can be faster but may sacrifice a little in resolution. Hence there is another method for high resolution microscopy, *Structured Illumination Microscopy (SIM)*.

5.4.3 *Structured Illumination Microscopy (SIM)*

One of the most elegant approaches to ultrahigh resolution optical microscopy is the use of structured illumination as a tool. The basic idea of this method, as presented by Gustafsson (2000) is that if one has two nearly equivalent set of finely spaced lines and we allow these two distinct grids to overlap, the pattern will exhibit a much more coarsely distributed set of grids in the overlap region. If the broad overlap region is then able to be clearly resolved, mathematically, one can perform a deconvolution on this pattern, and knowing one of the grid's pattern precisely, one can calculate the other pattern in a precise manner (**Figure 5.22**).

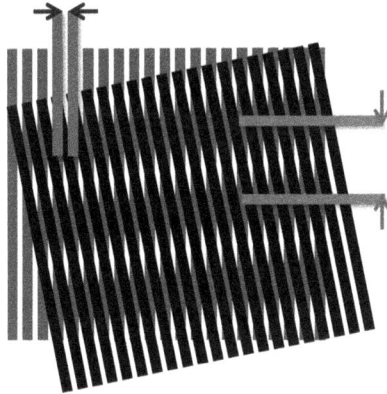

Figure 5.22. Sample features that are smaller than the wavelength of light (shown in gray grid) become the longer wavelength spatial "beat" frequencies when it is superimposed by another finely spaced grid of known periodicity (black). The latter is the concept of structured illumination.

In this example, consider the gray vertical grid pattern to be the sample of interest. Let the black grid mask pattern be the imposed known grid pattern; then the intersections form a more coarsely spaced pattern, yielding a pattern that is easier to resolve with accuracy. These are called the Moire fringes. Now if the sample of real interest has a random distribution of shape in space, we can consider that to be the spatial Fourier contribution from a whole set of well-defined patterns, when summed, will yield the image. Consider next the illumination of this object by a known grid pattern (like the blue one). Rotating this grid pattern maps out the contribution of Moire intensities by the presence of the sample, but because the Moire grids are easily resolvable, their patterns are easily calculated. The composite of many Moire pattern intensities allows for Fourier transform representation of the sample image. The resolution limit is usually $\frac{1}{2}$ the sampling grid's spacing. The advantage of SIM over STED is that this is not a point sampling scheme, thus allowing the entire image to be sampled in one sampling time frame. Furthermore, the field of view covered by SIM can be very broad, compared to PALM or STORM. In the most elementary application, SIM can improve the spatial resolution to about 100 nm. Nonlinear SIM techniques are being explored for higher harmonic gridlines, hence bettering the spatial resolution. This is still in research stage. The initial SIM images, however, required many sampling grid rotations for total image reconstruction, thus making it a slower processing scheme.

In a set of images provided by Thomas Huser (Monkemoller *et al.*, 2014), comparisons of the advantages of SIM and d-STORM are clearly seen. **Figure 5.23** shows the many complex components of a rat liver sinusoidal endothelial cell (LSEC). The main image is that of a multi-fluorophore labeled cell obtained using 3D-SIM. We note the broad span of the image coverage, with multi-color and hence in one image, know the approximate locations of many labeled components (see caption). This is the advantage of SIM technique. In going a little more in depth, images shown in panel upper right are low and high resolution representations of the tubulin domain using SIM. Shown in the far-right lower panel is the same tubulin images taken using d-STORM. Note the higher spatial resolution as evidenced by the more precisely located dots of reconstructed fluorophore centroids. The spatial resolution differential is given in a graph associated with these images. So d-STORM has the potential of achieving higher spatial resolution than SIM, as configured, but at an expense. The single frame exposure time of the d-STORM image was 20 ms and a total of 10,000

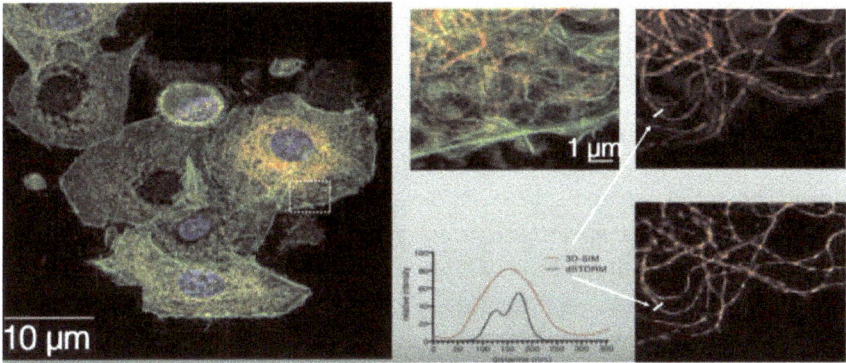

Figure 5.23. Large panel shows three-dimensional-SIM image of LSECs in multi-labeled fashion. Nucleus is in purple, actin network in green. Tubulin network is shown in red. The SIM image has ~ 100 nm spatial resolution. 10× magnified image of a boxed domain shown in middle top panel. The tubulin network alone is shown on top right panel. Using d-STORM, the right lower panel is achieved for the same domain of tubulin network. The higher spatial resolution is indicated by the more focused network using d-STORM method. Resolution limit of both techniques is given in graph. Figure courtesy of Huser (personal communications).

frames were used for the reconstruction. In these studies, SIM is used for initial, broader scans while d-STORM is employed for detailed studies of limited regions.

Extending SIM into the direction of more rapid sampling has the advantage of lower excitation level so as not to damage the biological samples, a more rapid sampling of the broad field, and still have, by using nonlinear saturation of the excited states, something like 50 nm spatial resolution.

Lattice and sheet microscopy methods are currently being advanced based on this concept (Chen *et al.*, 2014). In this approach, excitation is created by a thin sheet of illumination light, striking the sample at 90° from the high NA objective used for fluorescence detection. Conceptually, if the entire sheet of a sample is illuminated at the same time, one can image the entire field and collect all the information from the illuminated sheet at the same time, subject only to the limitation of the camera capture rate. In such an effective parallel sampling configuration, the amount of light needed for illumination is reduced, decreasing the chance of photobleaching or photo-induced damage to the sample without a decrease in signal to noise ratio (S/N). The added capability of creating sheets into lattices then allow for depth sampling in a controlled manner, adding the feature of optical sectioning capability. A comprehensive review of this topic is published by Power and Huisken (2017).

5.5   Multiphoton Excitation Fluorescence

Given the rapid advances in laser development in the late 20th century, applications to laser spectroscopy extended into the domain of nonlinear optics. As we have alluded to in Chapter 2, when the intensity of excitation of a light source becomes very high, materials respond in nonlinear manners. In such manner, subjects such as second harmonic generation, sum frequency generation and four-wave mixing evolved. These have very useful applications in biology, and we will devote Chapter 6 to discuss those topics as part of the label-free approaches. Here, we discuss one of the simplest applications of nonlinear optics: multiphoton excited emission (MPE) emission processes, such as fluorescence.

One issue that we mentioned in this chapter is the ability to actually excite the reporter groups of biological molecules, here, the fluorophore. Due to the optically dense matter that exists in the cell, often multiple scattering and inadvertent absorption of incident light decreases the amount of light flux that reaches the reporter fluorophore, hence the efficiency of the fluorophore is decreased. To overcome this limitation, we tap into a well-known feature of light: since the intensity of scattered light diminishes with increasing excitation wavelength, I_{sc} is proportional to $1/\lambda^4$, by increasing the wavelength of the reporter excitation the probability of multiple scattering will decrease. Here, we encounter a very fortunate situation: common dye molecules used as reporting fluorophores are typically absorbing in the visible. However, it takes two or more infrared photons to be able to excite that fluorophore. Since infrared light indeed scatters less than the visible or UV light wavelengths by our above discussion, we have reduced the depth of penetration issue. This developed into multiphoton excitation fluorescence (MPEF). The idea is quite simple.

Consider the material's first electronic excitation level to be at energy, E_1 that is not reachable by a single photon excitation $\hbar\omega$ from the ground state E_0. That is

$$\hbar\omega \leq E_1 - E_0 .\tag{5.10}$$

However, due to the energy breadth of the excited state, having come from a molecular system with many vibrational states associated with its existence, we see that two (or more) photons of energy $\hbar\omega$ will be able to bring the photon energy higher than the energy gap. Thus, we see that

$$2\hbar\omega \sim E_1 - E_0 . \qquad (5.11)$$

In this case, we call it two-photon excitation. So if the excited state, E_1, is fluorescent, the fluorescence created by this can be called two-photon excited (TPE) fluorescence. The energetics can be sketched as follows.

Illustrated in **Figure 5.24**, we note that the situation on the right (b), corresponds to our TPE fluorescence situation, but the one illustrated on the left (a), is the equivalent single photon excited fluorescence. That means the states of the material can be excited by either a single photon of energy (blue) or two photons (red). As depicted in this illustration, the red photons are only half as energetic as the blue.

One asks the question if laser sources at either blue or red were available, why would one choose one over the other? True, energetically they elicit the same response function, the green fluorescence signal. However, because the

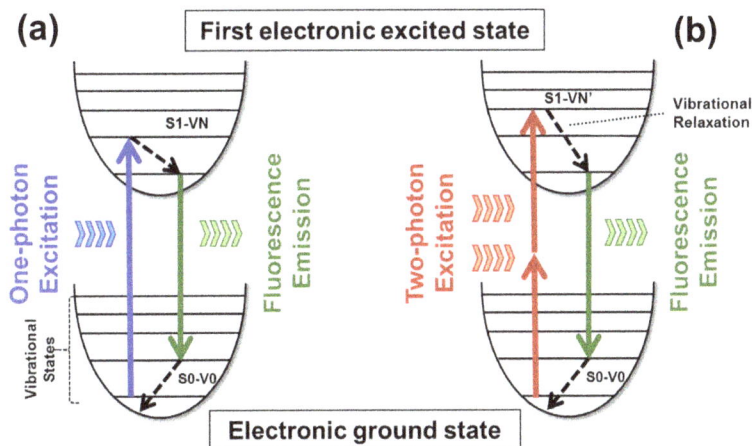

Figure 5.24. Jablonski diagram for one-photon (left) and two-photon (right) excitation. One-photon excitation occurs through the absorption of a single photon (blue arrow). The initial (S0-V0) and final (S1-VN) states have opposite parity. Two-photon excitation occurs through the absorption of two lower energy photons (two red arrows) via short-lived intermediate states. Note that arrow length denotes photon energy, not wavelength. In the two-photon case the initial (S0-V0) and final (S1-VN0) states have the same parity. After either excitation process, the fluorophore relaxes to the lowest energy level of the first excited electronic states via vibrational processes. The subsequent fluorescence emission processes for both relaxation modes are the same. Adopted from So *et al.* (2000).

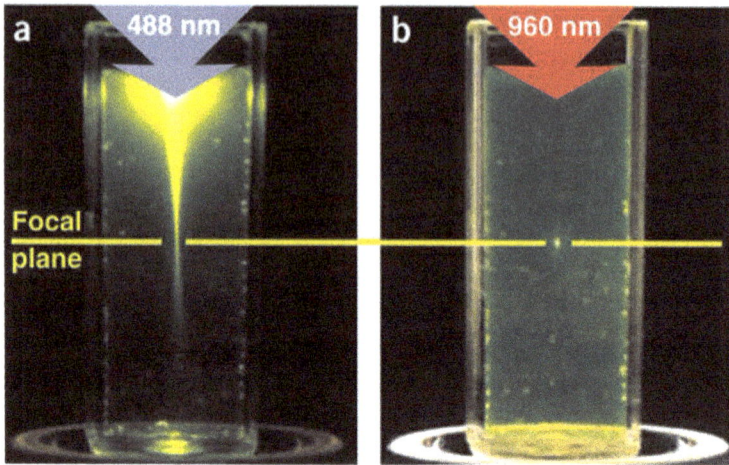

Figure 5.25.   Localization of excitation by two-photon excitation. (a) Single-photon excitation of fluorescein by focused 488-nm light (0.16 NA). (b) Two-photon excitation using focused (0.16 NA) femtosecond pulses of 960-nm light. Reproduced with permission from Zipfel *et al.* (2003).

red photons require a nonlinear response from the material, it comes about only when the applied field strength is high enough to elicit that response, rendering it a more selective process than using the blue photon.

In the illustrated experiment (**Figure 5.25**), a cuvette has a dye solution of fluorescein that fluoresces in the green. The left image shows 488 nm light focused into the solution and yellow fluorescence emission along the entire path of laser traversal. In the right image, 960 nm light is used to excite the same dye solution. Only in the center of focus (focal plane) of the laser beam do we see the excitation of the fluorescin dye molecules. This illustrates the ability of the two-photon excitation method to spatially select the domain of excitation, particularly in the depth sense. This is very similar to confocal imaging, leading to a technique called *optical sectioning*.

This approach has been widely used for fluorescence excitation in very complex space, such as the interior of the cell, particularly with the advent of very high numerical aperture (NA) microscope objectives. An example of this is shown below:

In **Figure 5.26** from Zipfel *et al.* (2003), 780 nm light was able to excite three different fluorophores within the RBL-2H3 cells. DAPI is used to label the nucleus, leading to the blue color, red depicts the mitochon-

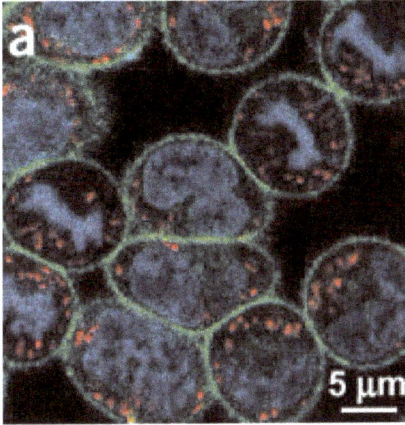

Figure 5.26.   Application of multiphoton microscopy. Simultaneous 780-nm excitation of three different fluorophores in RBL-2H3 cells labeled with 4′,6-diamidino-2-phenylindole (DAPI) (DNA, blue pseudo color), PATMAN (plasma membrane, green) and tetra-methylrhodamine (mitochondria, red). Reproduced with permission from Zipfel *et al.* (2003).

dria labeled by tetramethylrhodamine, and the green is a fluorescin-labeled sphingomyelin membrane component. The ability to conduct selective excitation using different light sources and TPE fluorescence makes this type of images possible.

So this is definitely advantageous! What can go wrong in such an experiment? In focusing very tightly, it is possible that the light field is so strong that not only do we have two-photon excitation as desired, we may be multiple-photon exciting the other regions of the molecule. This may lead to local excitation into higher excited electronic states with unknown de-excitation process, such as heating. In the extreme situation, MPE can lead to ionization and total destruction of the fluorophore. Autofluorescence from non-specified domains can create background noise that is difficult to eliminate. So care must be exerted in conducting these experiments. An aspect of using phosphorescence or luminescence radiation instead of the rapidly "firing" fluorescence may be an answer. Since these signals last over long time periods, one simply does not turn on the detector until the rapid, unwanted autofluorescence signals have disappeared over the course of a few nanoseconds. Such "time gating" ideas have been explored by many research groups.

5.6   Summary

A summary of the results of spatial localization efforts in the last decade is shown in **Figure 5.27** (Chiu and Leake, 2011; Fernandez-Suarez and Ting, 2008; Schermelleh *et al.*, 2010). As can be seen, progress is continuously being made in many of these areas. A further consideration across all of these techniques shown in this figure place an emphasis that the techniques aimed at increasing spatial resolution must also deal with the fact that

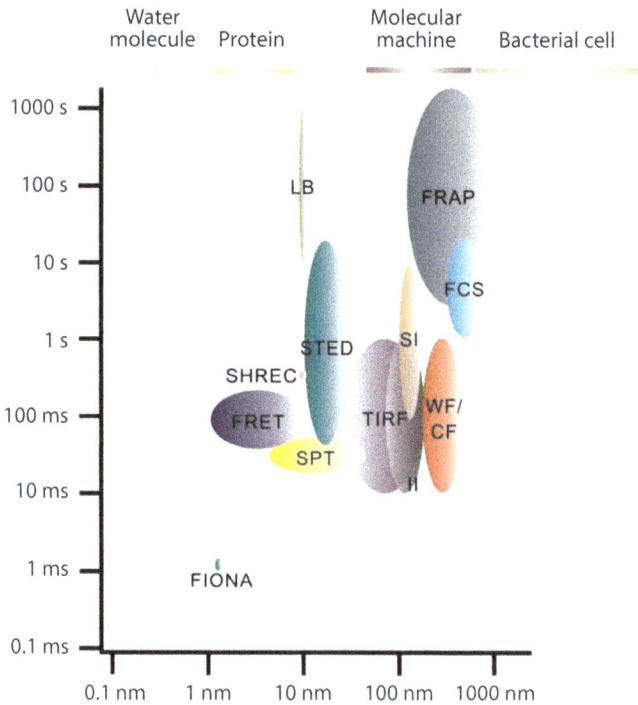

Figure 5.27. Comparison of the spatial and temporal resolutions of fluorescence microscopy. The length and time scales are logarithmic. The vertical scales refer to the amount of time needed to take one image frame or complete one measurement, the reciprocal of which represents the maximum rate at which dynamic changes in the sample can be detected. FCS (fluorescence correlation spectroscopy); FIONA (fluorescence imaging with one nanometer accuracy); FRAP (fluorescence recovery after photobleaching); FRET (Förster resonance energy transfer); II (interference illumination); LB (localization-based); SHREC (single-molecule high-resolution colocalization); SI (structured-illumination); SPT (single particle tracking); STED (stimulated emission depletion); TIRF (total internal reflection fluorescence); WF/CF (wide-field/confocal). Reproduced under Creative Commons Attribution License from Chiu and Leake (2011).

biological molecules are often very dynamic. Hence, time resolution (vertical scale of this figure) is another necessary consideration. We will delve into that aspect in Chapter 7.

References

Basil, A. and W. Wassef. Confocal endomicroscopy. In: N. Lagali (ed.) *Confocal Laser Microscopy – Principles and Applications in Medicine, Biology, and the Food Sciences.* Rijeka, InTech, 2013.

Betzig, E., G.H. Patterson, R. Sougrat, O.W. Lindwasser, S. Olenych, J.S. Bonifacino, M.W. Davidson, J. Lippincott-Schwartz and H.F. Hess. Imaging intracellular fluorescent proteins at nanometer resolution. *Science* 313: 1642–1645, 2006.

Chen, B.C., W.R. Legant, K. Wang, L. Shao, D.E. Milkie, M.W. Davidson, C. Janetopoulos, X.S. Wu, J.A. Hammer, 3rd, Z. Liu, B.P. English, Y. Mimori-Kiyosue, D.P. Romero, A.T. Ritter, J. Lippincott-Schwartz, L. Fritz-Laylin, R.D. Mullins, D.M. Mitchell, J.N. Bembenek, A.C. Reymann, R. Bohme, S.W. Grill, J.T. Wang, G. Seydoux, U.S. Tulu, D.P. Kiehart and E. Betzig. Lattice light-sheet microscopy: Imaging molecules to embryos at high spatiotemporal resolution. *Science* 346: 1257998, 2014.

Chiu, S.-W. and M.C. Leake. Functioning nanomachines seen in real-time in living bacteria using single-molecule and super-resolution fluorescence imaging. *Int. J. Mol. Sci.* 12: 2518, 2011.

Da Silva, L.B., J.E. Trebes, R. Balhorn, S. Mrowka, E. Anderson, D.T. Attwood, T.W. Barbee, Jr., J. Brase, M. Corzett, J. Gray, J.A. Koch, C. Lee, D. Kern, R.A. London, B.J. MacGowan, D.L. Matthews and G. Stone. X-ray laser microscopy of rat sperm nuclei. *Science* 258: 269–271, 1992.

Dale, B.M., G.P. McNerney, W. Hübner, T.R. Huser and B.K. Chen. Tracking and quantitation of fluorescent HIV during cell-to-cell transmission. *Methods* 53: 20–26, 2011.

Delano, W.L. PyMOL. The PyMOL Molecular Graphics System, Version 1.2r3pre, Schrödinger LLC, 2002.

Fernandez-Suarez, M. and A.Y. Ting. Fluorescent probes for super-resolution imaging in living cells. *Nat. Rev. Mol. Cell. Biol.* 9: 929–943, 2008.

Gustafsson, M.G.L. Surpassing the lateral resolution limit by a factor of two using structured illumination microscopy. *J. Microsc.* 198: 82–87, 2000.

Hell, S.W. Far-field optical nanoscopy. *Science* 316: 1153–1158, 2007.

Hell, S.W., M. Dyba and S. Jakobs. Concepts for nanoscale resolution in fluorescence microscopy. *Curr. Opin. Neurobiol.* 14: 599–609, 2004.

Hübner, W., G.P. McNerney, P. Chen, B.M. Dale, R.E. Gordon, F.Y.S. Chuang, X.D. Li, D.M. Asmuth, T. Huser and B.K. Chen. Quantitative 3D video microscopy of HIV transfer across T cell virological synapses. *Science* 323: 1743–1747, 2009.

Lin, Y.-H. and D.P. Tsai. Near-field scanning optical microscopy using a super-resolution cover glass slip. *Opt. Express* 20: 16205–16211, 2012.

Moerner, W.E. Polymer luminescence – Those blinking single molecules. *Science* 277: 1059–1060, 1997.

Monkemoller, V., M. Schuttpelz, P. Mccourt, K. Sorensen, B. Smedsrod and T. Huser. Imaging fenestrations in liver sinusoidal endothelial cells by optical localization microscopy. *Phys. Chem. Chem. Phys.* 16: 12576–12581, 2014.

Pedrotti, F.L. S.J., S.L. Pedrotti, M.L. Pedrotti. *Introduction to Optics*, Upper Saddle, NJ, Pearson Prentice Hall, 2007.

Power, R.M. and J. Huisken. A guide to light-sheet fluorescence microscopy for multiscale imaging. *Nat. Methods* 14: 360–373, 2017.

Rust, M.J., M. Bates and X. Zhuang. Sub-diffraction-limit imaging by stochastic optical reconstruction microscopy (STORM). *Nat. Methods* 3: 793–795, 2006.

Sauer, M. Localization microscopy coming of age: From concepts to biological impact. *J. Cell Sci.* 126: 3505–3513, 2013.

Schermelleh, L., R. Heintzmann and H. Leonhardt. A guide to super-resolution fluorescence microscopy. *J. Cell Biol.* 190: 165–175, 2010.

Shroff, H., C.G. Galbraith, J.A. Galbraith, H. White, J. Gillette, S. Olenych, M.W. Davidson and E. Betzig. Dual-color superresolution imaging of genetically expressed probes within individual adhesion complexes. *Proc. Natl. Acad. Sci.* 104: 20308–20313, 2007.

Sisan, D.R., R. Arevalo, C. Graves, R. Mcallister and J.S. Urbach. Spatially resolved fluorescence correlation spectroscopy using a spinning disk confocal microscope. *Biophys. J.* 91: 4241–4252, 2006.

So, P.T.C., C.Y. Dong, B.R. Masters and K.M. Berland. Two-photon excitation fluorescence microscopy. *Annu. Rev. Biomed. Eng.* 2: 399–429, 2000.

Trautman, J.K., J.J. Macklin, L.E. Brus and E. Betzig. Near-field spectroscopy of single molecules at room-temperature. *Nature* 369: 40–42, 1994.

Tsai, D.P. and Y.H. Lin. Super-resolution optical microscopy using a near-field cover glass slip. *SPIE Newsroom*, 2012.

Vlasov, I.I., A.A. Shiryaev, T. Rendler, S. Steinert, S.Y. Lee, D. Antonov, M. Vörös, F. Jelezko, A.V. Fisenko, L.F. Semjonova, J. Biskupek, U. Kaiser, O.I. Lebedev, I. Sildos, P.R. Hemmer, V.I. Konov, A. Gali and J. Wrachtrup. Molecular-sized fluorescent nanodiamonds. *Nature Nanotechnology* 9: 54–58, 2014. DOI:10.1038/nnano.2013.255.

Zipfel, W.R., R.M. Williams and W.W. Webb. Nonlinear magic: Multiphoton microscopy in the biosciences. *Nat. Biotechnol.* 21: 1369–1377, 2003.

Chapter 6

Label-free High-resolution Microscopy

A common feature of the advances in ultrahigh resolution optical microscopy discussed in Chapter 5 involves the use of exogenous labels such as a fluorophore. Though the advances in fluorescence-based microscopy have been in the spotlight, there are some major concerns whenever exogenous labels are used.

(1) The fluorescent labels attached to the molecule of interest may become a detriment in many ways. The typically large size of the labeling fluorophore, including the GFP, when compared with the biomolecule of interest, could be affecting the dynamics of the molecule of interest. A change in these dynamics could alter the kinetics of essential biochemical reactions.

(2) A related concern is the need to excite these fluorophores repeatedly for extended periods of time (as in PALM method) or with rather high-intensity pulses of light (as in STED). Photobleaching could become a problem if the fluorophores are no longer in the emitting state in a controllable manner. Repeated excitation could also create unwanted radiation damage to the spatial domain in the vicinity of the fluorophores, including the molecule of interest.

(3) The use of smaller reporters, such as quantum dots or rare earth nanoparticles have been known to be toxic to the biological molecules within the biochemical pathways. There needs to be shielding of these potentially toxic elements. This is usually done by encapsulating the reporting fluorophore or nanoparticle in a shell of lipid, via self-assembled monolayer (SAM) protocol. This procedure is still in a state of needing refinement. The current development of rare earth emitters within diamond structures as small as 5 nm, called diamondoids, may become a significant new approach to solve this problem.

Because of these concerns, the optics community has placed much focus on finding ways to not use exogenous reporters for locating or monitoring significant biomolecular structures and their dynamics. The two avenues currently explored are the use of nonlinear methods of excitation and the use of a shorter wavelength light source (e.g., X-rays) that will directly interrogate the molecular structures of interest.

6.1 Many Faces of Nonlinear Interaction

From our description of the fundamental tenets of nonlinear interaction, we noted that this is an effect that is derived from the nonlinear response of the materials when an incident electric field has been applied to them. In order to elicit any nonlinear response function, the material's polarizability has to be extended to higher orders of its response function, as we had developed in Chapter 2. Upon excitation by an incident field, $\mathbf{E}$, the macroscopic polarization $\mathbf{P}$, responds in the manner:

$$\mathbf{P} = \varepsilon_0 \left[\chi_1 \mathbf{E} + \chi_{NL} \mathbf{E}^{(>1)} \right]. \tag{6.1}$$

We see that $\mathbf{P}$ is now a nonlinear (NL) equation with respect to the incident field, $\mathbf{E}$, by virtue of its nonlinear dependence on this field. Since the nonlinear response function χ_{NL} is generally a very small quantity compared when with χ_1, only when the incident field is of sufficient magnitude will the second term of Eq. (6.1) be large enough to make a significant contribution to the total polarization. With the use of a laser as a light source, using its spatial and temporal coherence properties, and essentially "piggybacking" into the technologies of the modern microscope, it has become possible to render the nonlinear contribution of the polarizability to be a significant part of the total imaging capability. Thus we need to explore the ever-extending list of contributions of nonlinear interactions to the field of biophotonics.

6.1.1 *Second harmonic generation (SHG) and imaging*

Second harmonic generation (SHG) is the very first nonlinear phenomenon that was realized after the invention of the laser (Bloembergen, 1965). This process has been thoroughly studied and many devices utilizing the SHG method were developed. Laser manufacturing companies routinely exploit SHG as a method to create new sets of coherent light sources. Application of this process to biological systems was a bit more complex. Because the material under investigation is small by macroscopic crystalline standards, the length of material (molecules within proteins or nucleic acids) that is capable of SHG is typically very small (nm). Examining the equation for the production of a strong SHG signal from Chapter 2, we find that the factor

$$\sin^2\left(\frac{\Delta k L}{2}\right)\Big/\left(\frac{\Delta k L}{2}\right)^2 \approx 1.0 \tag{6.2}$$

for almost any phase matching Δk conditions because L, the length of the material traversed by incident light, is so small. This expression suggests that the macroscopic phase matching condition is now not the key criterion for the existence of SHG generation signal in biological materials. Instead, the signal will be small in all situations because L is short. That is, the amount of material that can create the nonlinear effect is limited. Thus the small L value renders SHG a difficult technique to use in the presence of other interfering features of biological matter, such as scattering.

In order to distinguish weak SHG signals from all others in a non-labeled environment, one relies upon another key criterion for the existence of SHG signal: a large nonlinear susceptibility, hence a non-centrosymmetric system. What this requirement specifies is that domains of uniqueness for the existence of the SHG signals, particularly for biological molecules, must have the physical property of non-centrosymmetry. Surface adsorbates and helical molecular conformations are among the most prominent. Surfaces or interfaces are intrinsically non-centrosymmetric with respect to the direction normal to the interface. Naturally, any small amount of adsorbate on a surface will create a noticeable change in SHG signal compared to that of the bare surface. Such signals have been obtained from protein adsorbates on crystal-solution interfaces (Brown *et al.*, 1985), from lipid membranes with unique protein receptors, and from surfaces specifically prepared for label-free receptor-antigen reactions (Wei and Shen, 2002). Shen's group (Wei *et al.*, 2002) had advanced the technique to investigate the structure of water near ice surface.

As for the helical systems, certain macroscopic protein systems form tissues that maintain a high degree of optical helicity or anisotropy. These are found in collagen-rich domains of tendons and ligaments. Striated muscle fibers also exhibit a substantial degree of optical anisotropy (recall Chapter 4). Many striking images have been obtained from SHG studies of collagenous tissues. We illustrate a couple.

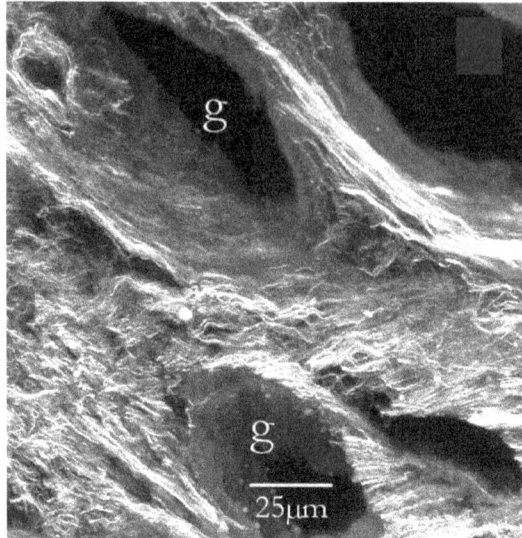

Figure 6.1.   Projection of three-dimension second harmonic generation. Unfixed human endometrium 50-μm cryosection, showing the pattern of collagen fibrils around a mucus gland (g). Excitation at 800 nm. 15° projection angle. A false-color palette has been applied. Reproduced with permission from Cox *et al.* (2003).

In the first image, shown in **Figure 6.1**, the human endometrium contains many collagen fibrils around a mucous gland (labeled g domains) as shown Cox *et al.* (2003). Because collagen is a triple helical polymer and is highly optically anisotropic, the SHG signal is strong. The image, though not at high resolution, is based on the existence of only these collagen molecules.

In a recent methods report by the group of Paul Campagnola (Chen *et al.*, 2012), a detailed description of the construction and operation of a modern SHG imaging system for biomedical applications is presented (**Figure 6.2**). These investigators discussed not only the instrumentation but also the steps to improve the quality and the capability of the method. Two aspects mentioned in this manual are:

(1) The ability to achieve polarization sensitivity. This feature allows the investigator to identify the orientation and pitch of helical order within the sampled region. This method allows for sampling of the structural order of the reporting molecules that are producing the signal. The most obvious example is the helical order of fibers within muscle, tendons,

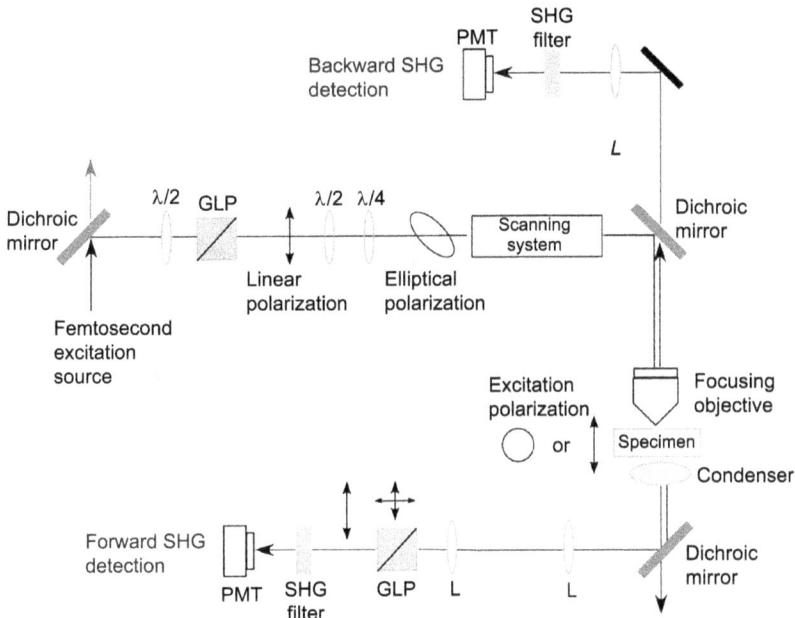

Figure 6.2.   Schematic of the optical layout of the SHG microscope, showing the optical components before the scan head and the detection pathways. Reproduced with permission from Chen *et al.* (2012).

and ligaments. It is possible to use polarization-sensitive SHG imaging to detect domains of helical order and disorder, even applying the latter method to discriminate a diseased or injured state of the tendon from that of the healthy. Stoller *et al.* (2002) developed the first polarizing SHG image method. Williams *et al.* (2005) added the criterion of differential orientation selection between the incident polarization and the SHG polarization signals. In their analysis, differential signals are shown at different polarization orientations for a single tendon. Thus SHG intensity profiles as a function of analyzer orientation are generated depending on the orientation of the incident polarizer. Shown in **Figure 6.3** in panels (c) and (d) are the SHG signals from the same tendon but at incident orientations of 0° and 90°. In this image comparison, the polarization orientation mapping shows how the difference in collagen helical arrangements is picked out by the polarization differentials. Such added features enhance the usefulness of optical imaging studies.

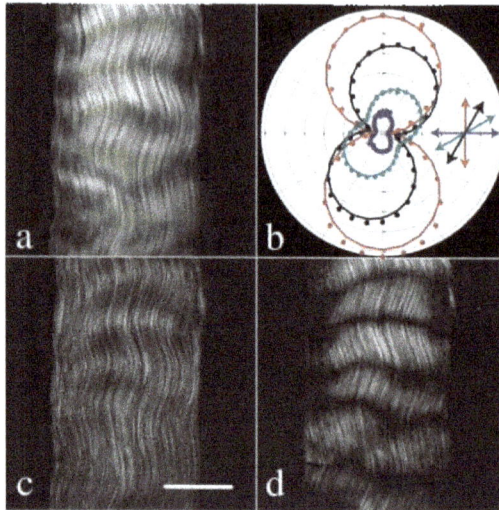

Figure 6.3. Tendon regions in which fibrils possess an orientation that is parallel to the fiber axis (a), yellow mask. In (b), the average SHG polarization intensity from this region is plotted with an illumination polarization at 0° (red), 30° (black), 60° (green), and 90° (blue) to the fiber axis (arrows). (c) shows the SHG intensity map of this tendon fiber, (d) is the same tendon with perpendicular polarization analyzers. Reproduced with permission from Williams *et al.* (2005).

(2) The ability to sample tissues of significant depth magnitude without exogenous labeling molecules. We have previously pointed out that non-linear methods have the same theoretical capability for optical sectioning as confocal microscopic methods. This is a very useful feature since tissues of any significant depth will have many different molecular compositions, each with a different index of refraction from the solvent medium, mainly water at $n = 1.333$. However, when incident light strikes such medium, light scattering takes place and competes against the ability to sample deeper regions. In order to achieve depth sampling, light scattering had to be minimized. The method used to achieve this end is called *optical clearing*, and was originally used in optical coherence tomography (OCT) (to be discussed in Chapter 8) studies where being able to sample deeper into the tissue is of major importance. In order to achieve optical clearing, an index matching fluid is used in place of the aqueous medium, rendering the index match between molecules of interest and that of the surrounding medium to be less prominent, hence less scattering. Often, glycerol is used for this function.

6.1.2 *Sum frequency generation — Possibly spectral imaging*

A variant of second-order nonlinear process is when instead of generating $\omega_{SH} = \omega_1 + \omega_1$, we have $\omega_{sum} = \omega_1 + \omega_2$, where the two frequencies ω_1 and ω_2 are different, even from different laser sources. From our analysis in Chapter 2, it is clear that the sum frequency generated (SFG) signal, being a second-order nonlinear response of the material, has the same phase matching condition for macroscopic samples: $\Delta \mathbf{k} = \mathbf{k}_{sum} - (\mathbf{k}_1 + \mathbf{k}_2) = 0$. For small domains such as a molecule where $L \sim 0$, again the phase matching criterion is automatically satisfied. Furthermore, there is still a non-centrosymmetry condition that must hold for SFG signals. The polarization field, P, now looks like

$$\mathbf{P}^{(2)}(\omega = \omega_1 + \omega_2) = \chi_{\text{eff}}^{(2)}(\omega = \omega_1 + \omega_2)\mathbf{E}_1(\omega_1)\mathbf{E}_2(\omega_2) \qquad (6.3)$$

where

$$\chi_{\text{eff}}^{(2)} = \chi_{\text{NR}}^{(2)} + \sum_q \frac{\chi_q}{\omega_2 - \omega_q - i\Gamma_q}. \qquad (6.4)$$

We note that in SFG, aside from a typically small non-resonant (NR) contribution, if the incident frequency of either of the two sources, here illustrated by using ω_2, is resonant with a material normal mode frequency, there will be an additional strong resonant signal. This allows the SFG method not only to sample the same domains exactly as SHG signals are derived, but a tunable ω_2 can sweep across the various molecular resonances. Typically, the SFG method is performed with one laser in the visible range while the other is an infrared (IR) sweeping light source. The material's fundamental vibrational frequencies will be resonant to the IR laser as it is tuned across them.

In **Figure 6.4**, we show schematically that at an intrinsically non-centro-symmetric region, a surface, a visible green laser source ω_1, combines with the IR red laser ω_2, to produce an SFG beam at a different wavelength, ω_3. An early study of molecular adsorbates on a surface was conducted by Follonier *et al.* (2003).

As described in Eq. (6.4), strong enhancement of the SFG signal at the detected sum frequency will correspond to the molecular resonances if the IR laser can be swept across the spectrum. This method is then advantageous for detecting the extent of specific molecular adsorbate onto a surface, with the measured polarization information indicating an orientational preference of the reporting molecular group of the adsorbates.

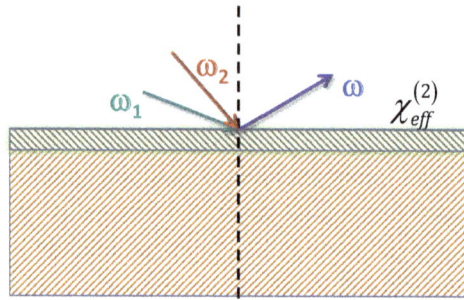

Figure 6.4.   Schematic representation of a SFG process.

The main idea is that the two distinct laser wavelengths, one visible and the second a tunable IR line, must arrive at the sample chamber synchronously. In a scheme recently developed by the Erlangen group (**Figure 6.5**), a high intensity pulsed visible light source was the output of a regenerative amplifier using a Ti-Sapp laser and a Nd:YLF pump laser. A small part of this visible beam is sent to the sample. The main part of this output is transmitted through an optical parametric amplifier followed by a difference frequency generator, rendering it a broadly tunable light source.

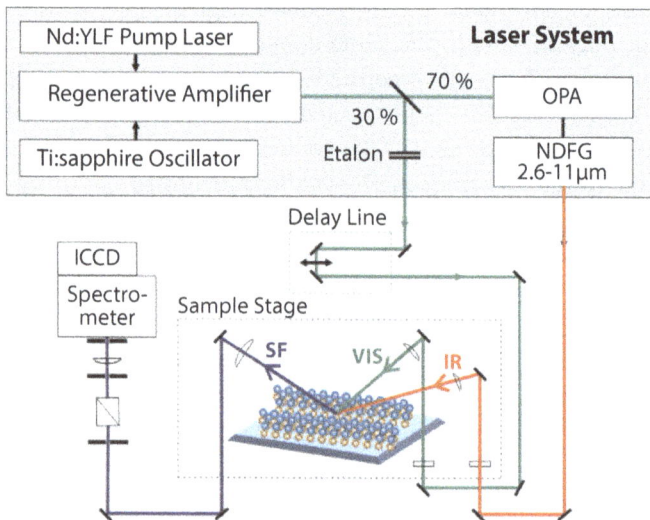

Figure 6.5.   Schematics of the setup for SFG approach. Image obtained through personal communication with Professor Peukert, University of Erlangen.

These two input lasers arrive onto the sample simultaneously by the use of an optical delay line, and the output SFG signal is then analyzed by a spectrometer and camera.

Using a laser scheme similar to the one described, Sohrabpour *et al.* (2016) has recently obtained vibrational SFG signal from C_{60} *fullerene* in the frequency range of 1400–1500 cm^{-1} (Figure 6.6). It is most interesting to note that the SFG signals are highly polarization sensitive and substrate sensitive, making the technique a beneficial one for probing unique biomolecular structures in different environments. In **Figure 6.6**, experimental vibrational SFG (VSFG) spectra of C_{60} on silica (red markers) and CaF$_2$ (blue markers) are shown in the two overlaid panels. In (a) the polarization orientation of the three fields in the ssp combination, where s designated field perpendicular to and p being in the plane of incidence, shows differential signal intensity at ~ 1470 cm^{-1} whereas in (b) sps polarization combination shows strong correlations between the two different substrates at 1430 cm^{-1}. Such spectral distinctions can render images with high specificity.

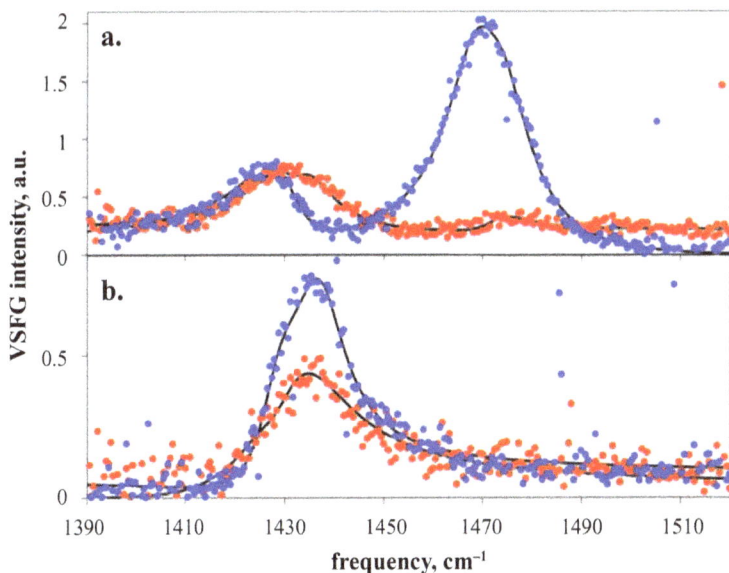

Figure 6.6. Experimental VSFG spectra of C_{60} on silica and CaF$_2$ overlaid with the multilayer interference fits (black line) for the (a) ssp polarization combination and (b) sps polarization combination. Reproduced with permission from Sohrabpour *et al.* (2016).

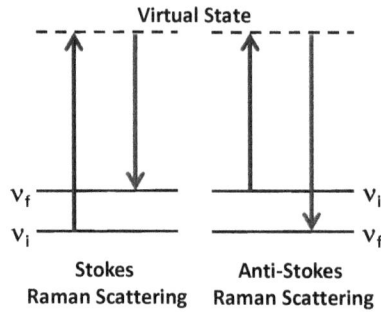

Figure 6.7.   Spontaneous Raman scattering with the downward arrows.

6.1.3   *Stimulated Raman Scattering*

Raman scattering process as discussed in Chapter 2 shows that for molecules that exhibit induced dipole moment (polarization) changes upon vibrational motion, there will be a modulation scattering signal at the vibrational frequency, ω_v. This is equivalent to saying that the polarization signal at the inducing optical frequency is now modulated by the normal modes of molecular vibrations. Thus, $\omega_R = \omega_0 \pm \omega_v$. Note that in this classical analysis, the modulation frequency should be detectable at either the Stokes side $(-)$ or the Anti-Stokes side $(+)$. However, if one examines the quantum energy levels, it becomes clear that the Anti-Stokes signal can be seen *only if the upper vibrational level is pre-populated*. Since there is a Boltzmann probability differential for occupancy of the upper versus the lower state, the predominant spontaneous Raman signal is generally from the Stokes scattering (**Figure 6.7**). Overall, because spontaneous Raman scattering is a second order process when compared with IR absorption, Raman scattering is a very weak signal. Surface Enhanced Raman Scattering (SERS) process can take advantage of plasmonic field enhancement if there are metallic surfaces adjacent to the molecule of interest. Gold or silver surfaces or nanoparticles have been used to achieve this field enhancement. This however, becomes almost a label hence the process is not considered label-free.

Stimulated Raman Scattering (SRS) process, however, renders Raman process a technique useful for biophotonic application. The idea behind SRS is that in this case, the nonlinear polarization is

$$\mathbf{P}_{\mathrm{NL}} = \varepsilon_0 \left[\chi_1 \mathbf{E} + \chi_3 \mathbf{E} \cdot \mathbf{E} \cdot \mathbf{E} \right] \tag{6.5}$$

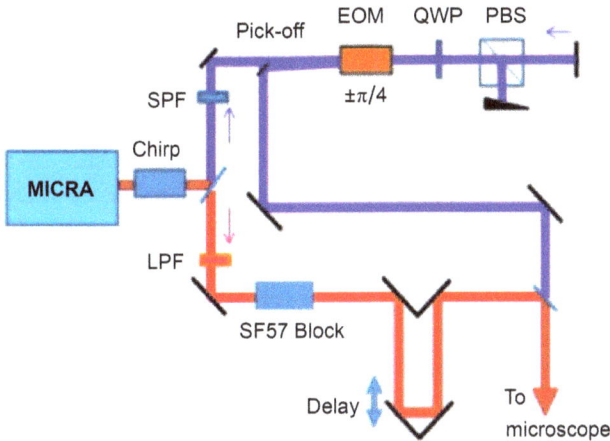

Figure 6.8.   Schematic diagram of the SRS spectroscopy setup. Short-pulse (fs) laser source is put through a frequency chirp device. The Stokes portion is passed through a frequency modulator scheme (blue). The pump source, chirp adjusted by SF57 block, is sent through an optical delay line that is adjusted to bring about simultaneous arrival of the Stokes and pump sources. Reproduced with permission from Fu *et al.* (2013).

where the dominant nonlinear susceptibility is χ_3. That is to say, the medium under excitation has a *center of symmetry*, hence the *second-order* nonlinear term is absent. Consequently the leading nonlinear term is of the *third-order* type. Let us examine the frequency of the nonlinear polarization field in the following manner: in normal spontaneous Raman (Stokes) scattering, $\omega_s = \omega_0 - \omega_v$. However, the vibrational frequency may be thought of as being produced by $\omega_v = \omega_0 - \omega_s$. Substituting, we have $\omega_s = \omega_0 - \omega_0 + \omega_s$, showing that the SRS process is a *degenerate Four Wave Mixing* process (**Figure 6.8**). The polarization field will be at the Stoke's frequency. This result indicates that the nonlinear susceptibility χ_3 exhibits the same vibrational resonances as in spontaneous Raman signals! Essentially, the process is one where the Stokes frequency light source achieves a signal gain upon entering such a nonlinear medium. Yariv's (1968) chapter on Stimulated Raman Scattering describes this topic (Chapter 2), which sometimes is referred to as Stimulated Raman Gain (SRG).

In an effort to discriminate the desired SRS signal from any background signal at this frequency, Xie's group developed a technique whereby one of the signals is modulated. In their original scheme (Freudiger *et al.*, 2008), either the pump or the Raman signal is amplitude modulated. The output is then put through a lock-in amplifier and detected at the modulation frequency. In such a configuration, the SRS signal is seen as a modulated gain signal on top of the existing Raman input, where the SRS modulation frequency is produced by the pump light at ω_0 modulated at the lock-in detected frequency. Utilizing this method for imaging basically requires the *a priori* identifying of a specific spectroscopic feature of a biological sample that will be useful for image presentation. In the initial work of the Xie group, the vibrational signal of the CH_2 groups of membranes at 2845 cm^{-1}, due to its high prevalence of lipid molecules, shows up very well compared with smaller quantities of proteins.

Advances in SRS has taken place primarily in the area of achieving multi-spectral imaging. Since biological systems all have complex molecular structures and spectra, it is most desirable to be able to ascertain the existence of a particular molecule by having its range of signature spectra exhibited. This means the ability to achieve *hyperspectral* SRS (hs-SRS) would be advantageous. In their original scheme, the pump and Stokes heterodyning process is essentially handling one Stokes frequency at a time. Hence the resulting image can display only one molecular bond type. In their more recent work, a frequency chirped femtosecond laser whose frequency spectrum includes the Stokes signatures of a particular molecule, be they membrane, nucleic acid or amino acids, is used as one of the laser sources. They have resorted to using frequency modulation (FM) as being more advantageous than the original amplitude modulation (AM) mode. In **Figure 6.9**, such hs-SRS spectrum is shown (Fu *et al.*, 2013).

Indeed characterizing the range of spectral signatures provides the investigator with much more confidence on the specificity of the molecules. In their latest work on monitoring the presence of acetylcholine (Fu *et al.*, 2017) shown in **Figure 6.10**, we can see in the image on the right that the signature 720 cm^{-1} is strong in the junction region after the fiber is subjected to electrical stimulation. The researchers caution that in this image, signals from the 720 cm^{-1} of phosphatidylcholine does also enter as a background due to its prevalence. Please consult the new volume edited by Cheng and Xie for extended details of this section (Cheng and Xie, 2017).

6.1.4   *Coherent Anti-Stokes Raman Scattering (CARS)*

A competitive third order nonlinear Raman process is the CARS. From our previous description of this anti-Stokes Raman scattering in the spontaneous sense, it is clear that this signal is a weak process when compared with spontaneous Stokes Raman scattering by the Boltzmann factor of

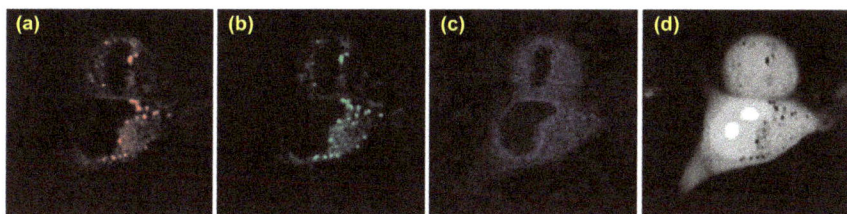

Figure 6.9.   Hyperspectral SRS. hr-SRS images of test domain obtained in a single setting. The pseudo colors show regions predominantly (a) saturated lipids DPPC, (b) unsaturated lipids DOPC, (c) cholesterols, and (d) proteins. Reproduced with permission from Fu *et al.* (2013).

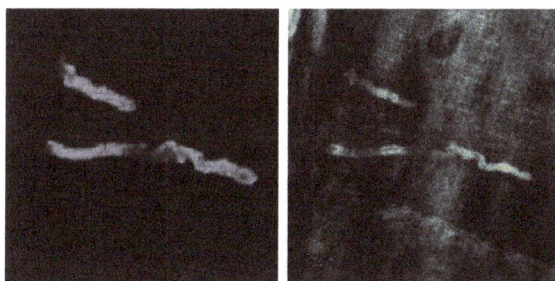

Figure 6.10.   Left: neuromuscular junction (NMJ) imaged using a-BTX staining. Right: FM-SRS of acetylcholine (ACH) signature ~ 720 cm^{-1} after junction excitation via electrical stimulation. The strong signal at the region of the junction indicates ACH release during electrical excitation. Background is from excess phosphatidylcholine membrane constituent. Reproduced with permission from Fu *et al.* (2017).

$\exp(-\Delta E/kT)$, where $\Delta E = E_1 - E_0$. Now, when considering the third order nonlinear process, we need to develop it as $\omega_{AS} = \omega_0 + \omega_v$. However, since $\omega_v = \omega_0 - \omega_s$, we can re-write $\omega_{AS} = 2\omega_0 - \omega_s$. What this says is that when energy conservation is established, we will achieve the third-order equivalent of resonant anti-Stokes Raman scattering. This is shown in the schematic diagram below. In this diagram, the double pump beams couple with a Stokes beam to produce the resultant CARS signal (**Figure 6.11**).

The fact that CARS is on the upside of the energy scale, it has the advantage of not being complicated with the often-present intrinsic autofluorescence signals in biological tissues. Thus it is often touted as a "cleaner" signal. The downside of this is the rather complex assembly of lasers in time and space so that the signal is produced when excitation beams are coincident in space and time. In **Figure 6.12**, the necessary

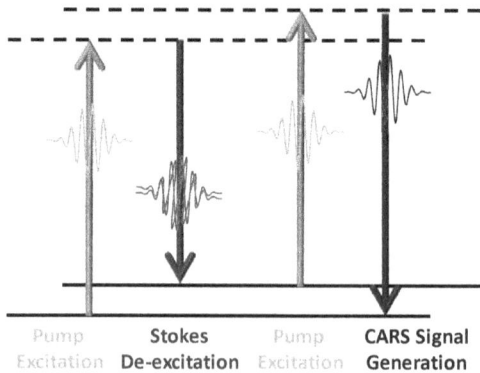

Figure 6.11. Schematic energetic description of the CARS signal: $\omega_{AS} = 2\omega_0 - \omega_s$.

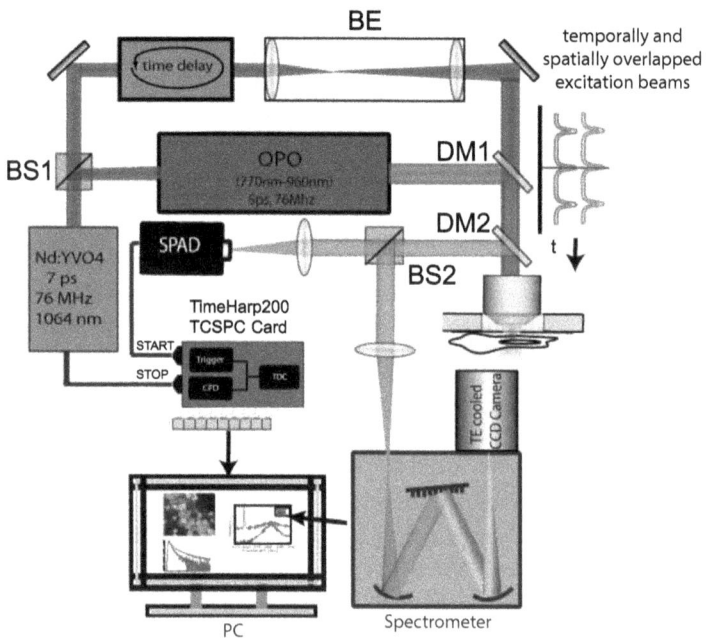

Figure 6.12. CARS microscopy experimental setup. An optical parametric oscillator is synchronously pumped by a Nd:YVO4 laser to produce the excitation beam, while the Nd:YVO4 laser provides the probe beam. CARS signals are detected in epi-mode by a single photon avalanche diode (SPAD). BS: beam splitter; DM: dichroic mirror; BE: beam expander; OPO: optical parametric oscillator; TCSPC: time-correlated single photon counting; PC: personal computer. Reproduced with permission from Ly *et al.* (2007).

time delay line and spatial beam expander needs are introduced in the pump beam at the excitation wavelength ω_0, in order to assure the proper simultaneous arrival of the two beams with nearly the same degree of spatial overlap of the beam from the Stokes wavelength ω_s. The single photon avalanche diode (SPAD) collects the anti-Stokes signal one spatial point at a time. The image is produced by raster scanning of the illuminated space.

Furthermore, there is a rigorous phase matching condition for the existence of CARS signals, given by

$$\mathbf{k}_{AS} = 2\mathbf{k}_0 - \mathbf{k}_S . \tag{6.6}$$

So depending on the specific Stokes frequency or wavevector, the anti-Stokes wavevector will be directed at a specific angle satisfied by the above equation. It is indeed possible to achieve the phase matching conditions both in the forward and backward directions. This provides two avenues for CARS detection, called Forward CARS or F-CARS and backward or Epi-CARS (E-CARS) to cover the backward scattering process. Microscopy applications based on F-CARS are often limited by a strong non-resonant background signal produced by the non-specific four-wave mixing process. E-CARS, on the other hand, is very sensitive to the size and shape of the sample and in large samples can be completely suppressed due to destructive interference. E-CARS also exhibits less non-resonant background contributions compared to F-CARS. In an experiment trying to capture the maximum amount of information with CARS configuration, Schie *et al.* (2008) designed their CARS experiment to use a single charge-coupled device (CCD) to capture both the F-CARS and E-CARS images. In this image, human Mesenchymal Stem Cells (hMSC) were cultured and induced to differentiate into fat cells. These MSCs begin for form lipid-rich droplets that permeate the entire cytoplasm. Since these lipids have aliphatic CH_2 bonds with symmetrical stretch vibration at 2845 cm^{-1}, this was the CARS signal that was used for image formation.

Images from these samples can be obtained both in the F-CARS and E-CARS. For this sample, they are shown in **Figure 6.13**. Here, F-CARS image is shown in panel a, E-CARS of the same sample is shown in panel b. The superposition of the two images is given in panel c, where the F-CARS signal is shown in red (false color), and E-CARS image is shown in green, both for the CH_2 vibrational bands of adipose cells. The white line in panel c defined a typical cross-sectional cut of the superposed images. These are

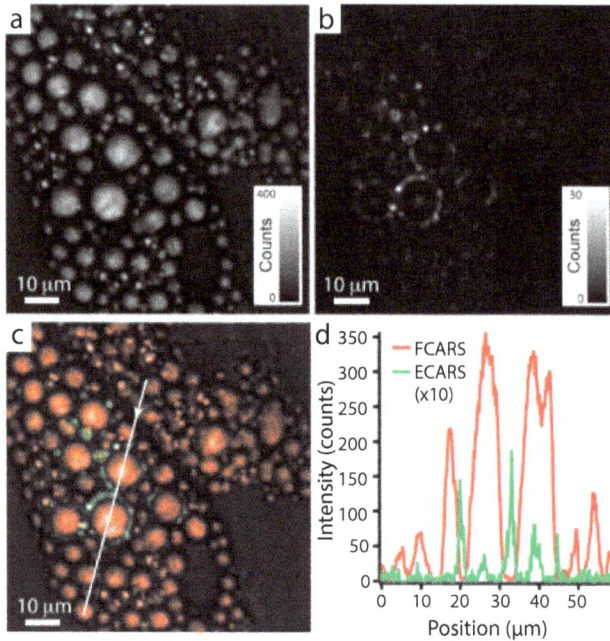

Figure 6.13. Time-gated CARS spectral images. (a) Time-gated intensity image of the F-CARS signal. (b) Time-gated E-CARS image of the adipocyte sample. (c) Overlay of the E-CARS and F-CARS image where contrast for the E-CARS image. (d) Line section of the CARS signals as indicated by the white line in (c). Reproduced with permission from Schie *et al.* (2008).

then exhibited as an intensity plot versus position in panel d. It is of interest to note that the two CARS signals complement each other in their image presentation.

Another interesting feature of CARS imaging is that because CARS is an instantaneous effect versus fluorescence which has typical lifetimes of 10^{-9} sec, a time-gating mode can be implemented to accentuate the immediate ($\ll 10^{-9}$ sec) response of the CARS signal, while the background of multiphoton excited fluorescence has not a chance to materialize yet ($\sim 10^{-9}$ sec).

Time-gated discrimination of background signals is an essential aspect of arriving at higher quality images. In **Figure 6.14**, Ly *et al.* (2007) show that the CARS signal in red (false color) accentuates local domains as lumens providing a signal at the lipid signal of 2945 cm^{-1}. The background from autofluorescence of the surrounding tissues are shown in green (false color).

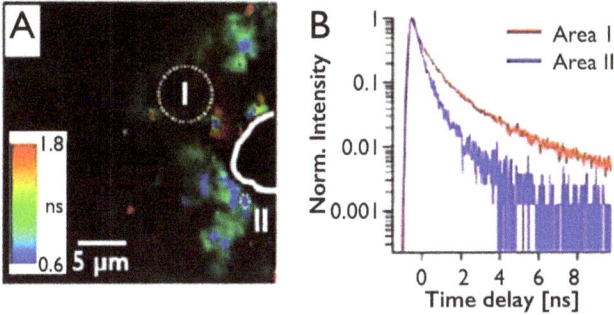

Figure 6.14. CARS micrograph of rat arterial tissue (cross-section) showing the effect of time-gating on animal tissue. (A) CARS image encoding the photon arrival time on a false color scale. Image scale is 30×30 μm. (B) Normalized photon arrival time histograms obtained by isolating areas with short photon arrival times from areas with long photon arrival times as highlighted by white circles in (A). Reproduced with permission from Ly *et al.* (2007).

6.1.5   *Self-focusing and self-phase modulation*

It is evident from the earlier discussion that the nonlinear effects are generated using high-intensity lasers. These are invariably derived from pulsed lasers, where the energy of the entire pump source in the laser is released by a sudden change in the condition of the laser cavity, either through Q-switching or mode-locking protocols, to create very short pulses (ps – fs variety). The third-order nonlinearity in polarization lends itself well to a set of self-induced effects that can prove to be either beneficial or detrimental to the experiment at hand. These are related to the term

$$\mathbf{P}_3^{\mathrm{NL}} = \varepsilon_0 \chi_3 \mathbf{E}^3 \,. \tag{6.7}$$

Since the three E fields can be one and the same field, we see that this polarization may be written as

$$\mathbf{P}_3^{\mathrm{NL}} = \varepsilon_0 \chi_3 I \mathbf{E} = K \mathbf{E} \tag{6.8}$$

where $K = \varepsilon_0 \chi_3 I$ is the dynamic Kerr coefficient, proportional to the square of the E field or the intensity, I. Since the polarization is now a function of the incident field intensity, it carries with it several interesting new observations.

(1) Self-focusing — As we know, this nonlinear part of the polarization is then related to the nonlinear part of the polarizability and nonlinear part of the index of refraction. This means that with third-order non-

linearity, using a Gaussian laser beam profile, the center of the profile has higher field than the edge. However, higher field leads to higher nonlinear index or higher overall index of refraction. That is to say, the index of refraction in the middle of the beam profile is higher than at the edges. Since $v = c/\eta$, the light wave travels slower in high index medium than in lesser index medium, the incident planar wave front bends towards the middle, essentially allowing the beam to focus. This self-induced effect is called *self-focusing*. In combination with the Airy diffraction spreading, stable self-focused filaments of light could form. Light filaments in a medium may be advantageous since it allows the radiant density to increase. Very early laser Raman scattering studies apparently had taken advantage of this self-focusing mechanism as an unknown helper for its enhanced S/N. With very high powered lasers, however, this self-focusing can lead to localized heating and multiphoton absorption in small spatial regions, causing the material to overheat, leading to defocusing (thermal blooming) or its material destruction.

(2) Self-phase modulation — The same process that we have described can also cause the phase of the traveling wave in a wavefront to differ as a function of the intensity of the temporal beam passes the material. The low-intensity initial beam entry point of a pulsed laser into the material may present a different nonlinear index of refraction than the peak of the pulse, which is of high intensity. This causes the index of refraction to be modulated as the pulse passes, leading to frequency *"chirping,"* which adds frequency components to the pulse. In many femtosecond pulse experiments, this phenomenon that happens in optical fibers is a severe limitation to the shortness of the amplified pulse output. Invariably a pulse shape restorer or compensator system needs to be introduced to recover the narrow pulse shape. This process, when uncontrolled, can lead to anomalous signals that cause difficulties in analysis of experiments. However, as we have seen in the development of SRS and CARS experiments for imaging of biological molecules, the presence of a band of frequencies allows for spectral focusing during the imaging process. It is of interest to point out that the use of femtosecond pulse for CARS spectroscopy was initially discussed by Scully *et al.* (2002).

(3) Multiphoton damages — Often we have used the high intensity and short pulse lasers to our advantage in the drive for using nonlinear optical means to probe biological molecules in a non-labeled environ-

ment. Beyond the two cautionary points above, we also need to mention *multiphoton absorption* damages. It is not an obvious one since we are not even talking about given levels of the material's electronic structure being able to accept the photons with one or two photon energy differences. Multiphoton absorption in an uncontrolled fashion may lead to unwanted heating of the sample and even ionization or vaporization of the matter at hand. Basically, the idea is to use as low a laser pulse power as possible to affect the end goal. Biological samples are often fragile against high-intensity pulsed lasers.

6.2 X-ray Microscopy

6.2.1 *Early soft X-ray microscopes*

Given the rather complex modes of optical microscopy in search for higher spatial resolution in biological sample imaging, one is tempted to approach the problem in a much more direct way. That means since the Airy disc formulation defines the point-spread function of a source point that is illuminated (either scattering or fluorescence), why not just use EM wavelengths that are shorter? Indeed, if one shifts to $\lambda = 1$ nm regime, with the assumed availability of focusing devices of NA ~ 1.0, we can get optical resolution in the 1 nm range! This was the idea of many early X-ray microscopy efforts (Da Silva *et al.*, 1992). What was envisioned was that organic, therefore biological materials, have many carbons, and there should be the carbon K-edge absorption that will absorb and be able to be differentiated from the surrounding aqueous medium if one were to capitalize on the transmission properties of soft X-rays from 2.3–4.4 nm (water window) regime. Because oxygen has an absorption K-edge at 2.34 nm, and carbon has its K-edge at 4.4 nm, the combination of these in this regime should provide clear absorption contrast between biological materials and the background. The Advanced Light Source (ALS) at Lawrence Berkeley National Laboratory had dedicated a beamline for the development of an X-ray microscope based on this principle (**Figure 6.15**).

The idea is to tap the synchrotron radiation from oscillating electrons in a beamline by deflecting that light into an X-ray reflector. The light can then be focused onto the sample using an X-ray zone-plate condenser lens. Using point-by-point sampling, the rastering of the sample stage will lead to a recording of optical signals from the presence of carbon (and oxygen)

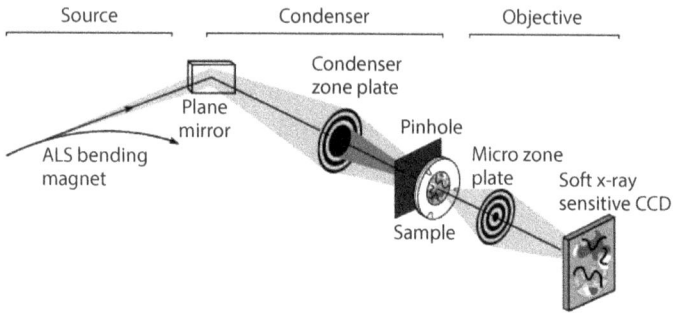

Figure 6.15. X-ray microscope XM-1. The microscope uses a microzone plate to project a full field image onto a CCD camera that is sensitive to soft X-rays. Partially coherent, hollow-cone illumination of the sample is provided by a condenser zone plate. A central stop and a pinhole provide monochromatization. Reproduced with permission from Chao *et al.* (2005).

rich items in the soft-X-ray window region. Since the PSF is now small, high resolution can be achieved in the neighborhood of 10 nm.

Difficulty nonetheless hampered the progress of the X-ray microscope using this principle. The main obstacles are the low fluence of X-rays and the damage created by the X-rays even at very low fluences. Samples that absorb these radiation wavelengths will cause ionization or even worse consequences on the biological materials. Soon it became apparent that to overcome these difficulties; the direct imaging approach must change its method completely.

6.2.2 *Coherent X-ray laser sources for microscopy*

A recent, more promising approach to the study of biological materials using X-rays is the development of the ultrashort pulsed hard < 0.1 nm X-rays in a coherent X-ray laser fashion. These are called X-ray free electron lasers (XFELs). The idea behind this approach is that instead of having a low dosage of 0.15 nm X-rays exciting a crystallized sample of a biological molecule and obtaining a diffraction pattern of the interaction, we now sample a single molecule of the needed biological significance in a beam of hard X-rays created by the ability of a new generation of XFELs. Instead of getting the crystalline diffraction pattern from macroscopic crystals, the diffraction pattern is derived from a stream of micro or nanocrystals, each smaller than the coherence length of the laser beam. The Bragg scattering signals from all the random orientations are captured across a wide range of angles through the use of a position-sensitive CCD camera. From these not-so-well defined patterns, the total inverse problem is carried out to obtain the Fourier transform of the diffraction pattern, which should be the image of the molecule.

Since the beam is striking the sample in a very short time, $\sim$ fsec time scale for the interaction time, the optical scattering or diffraction signal has to be captured in that short time before the hard X-rays damage the samples. Thus in order to build up sampling statistics, the nanocrystalline material needs to be fed into the coherent light beam in a synchronized repetitive fashion.

Such an idea has become a reality recently at the Linac Coherent Light Source (LCLS) at Stanford, California and Deutsches Elektronen-Synchrotron (DESY), Hamburg, Germany. The schematic instrumental system is shown in **Figure 6.16**.

In this, one of the earliest successful experiment, LCLS provides coherent light pulses of 1.8 keV X-rays (0.69 nm) 70 fs duration, and with 2.6×10^{12} photons/pulse. A stream of nanocrystals of Photosystem I was injected into the path of the LCLS X-ray laser pulse. The two fast CCD cameras collect both the wide angle X-ray scattering (WAXS) and small angle X-ray scattering (SAXS) data from the sample laser interaction. The number of samples sent into the path exceeds 3,000,000. The collected

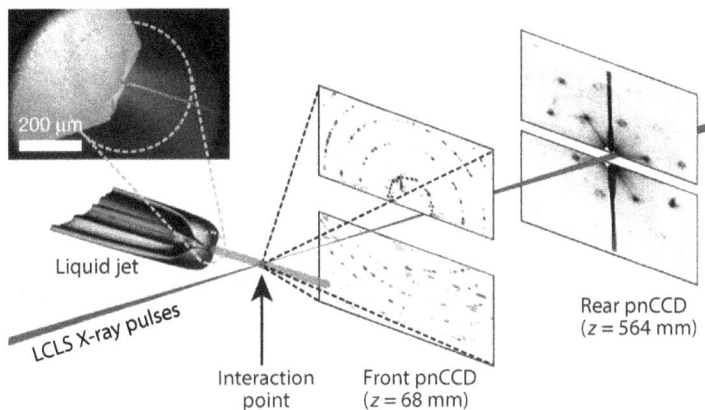

Figure 6.16. Femtosecond nanocrystallography. Nanocrystals flow in their buffer solution in a gas-focused beam that is focused on the jet perpendicular to the pulsed X-ray FEL. Two pairs of high-frame-rate pnCCD detectors record low- and high-angle diffraction from single X-ray FEL pulses. Reproduced with permission from Chapman *et al.* (2011).

diffraction patterns are subject to the phase sensitive inverse scattering calculation protocol that was perfected by Chapman *et al.* (2011). The initial study was performed on a membrane protein complex called Photosystem I (PSI). This system has a molecular weight of ~ 1 Mega Dalton (MDa), and the nanocrystals have two trimers per unit cell. Image reconstruction was achieved from many thousands of ~ 2 keV XFEL snapshots taken from many nanocrystals in random orientations. The resultant reconstruction of this PSI membrane protein structure is shown in **Figure 6.17**, where (a) shows the two diffraction patterns from the SAXS and WAXS data; (b) is the summed pattern from the tens of thousands of diffraction

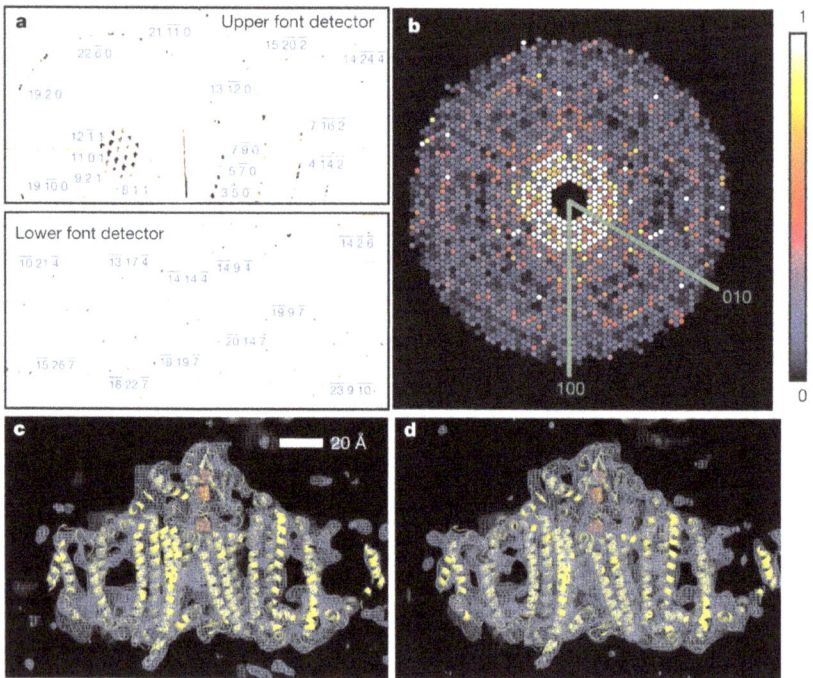

Figure 6.17.   Diffraction intensities and electron density of photosystem I. (a) Diffraction pattern recorded on the front pnCCDs with a single 70-fs with a resolution 8.5 Å. (b) Precession-style pattern of the [001] zone for photosystem I, obtained from merging femtosecond nanocrystal data from over 15,000 nanocrystal patterns, displayed on the linear color scale shown on the right. (c) and (d) Electron density map calculated from conventional synchrotron data truncated at a resolution of 8.5 Å and collected using femtosecond X-ray protein nanocrystallography. Reproduced with permission from Chapman *et al.* (2011).

patterns; and (c) and (d) show the reconstructed image of the protein structure. The protein system is shown in gray. The locations of the transmembrane helices are highlighted in the yellow ribbons. The defined orientation of those helices suggests that the reconstruction is indeed of a membrane protein in its relatively native state.

In a more recent study by Arnlund *et al.* (2014), the beam energy has reached 9 keV, allowing for spatial resolution of less than 0.2 nm. A report shows the ability of this method to capture changes in molecular structure. The system under study is the photosynthetic reaction center RC_{vir} of *Blastochloris viridis*. Being a photosynthesis center, the protein absorbs light and carries out its function. The authors wanted to show that upon capture of light at a localized domain of the protein, the protein structure actually undergoes a sudden redistribution of captured energy before carrying out the requisite function. This event, called "quaking," happens within a picosecond of the light capture. In order to monitor this motion, the authors prepared membrane vesicles that contained these reaction centers. Then a solution of these RC was delivered to the free electron beam in rapid pulses. In this case, jet streams flowing at 10 m/sec of solubilized samples were spray injected into the beam. While in the beam, the RC_{vir} reaction center is set into motion by directly exciting with an 800 nm pump laser source. The CCD captures the events as WAXS pattern that is changed in its distribution upon the light excitation (**Figure 6.18**).

Data was analyzed similarly but captured in the time sequence of the quaking events. Reconstructing the image, shown here in a stereo pair (**Figure 6.19**), the domains related to shaking or quaking are shown in yellow dots.

Capturing events of non-crystallized protein molecules in dynamic motion is definitely a *tour de force* with the present apparatus. It required the most detailed sample solution preparation, precise synchronization of the pulses (X-ray and laser) delivery, and the painstaking analysis of reams of data.

In a more recent work, Gati *et al.* (2017) describe the atomic resolution structure of native nanocrystalline granulovirus occlusion bodies (OBs) that are less than 16 nm^3 in volume using the full power of the LCLS with excitation dose up to 1.3 GGy per crystal (1 Gy (gray) = absorption of one joule of radiation energy per kilogram of matter). Each of these crystalline shells of granulovirus OBs, on the average, consists of about 9,000 unit cells. In the analysis conducted by this group, molecular bonding structures at

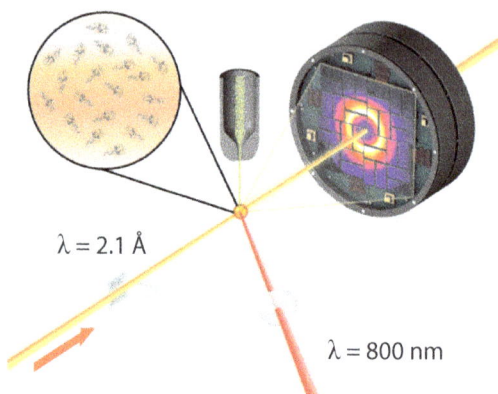

Figure 6.18.   Experimental setup illustrating the time-dependent changes in wide angle X-ray scattering data recorded from detergent-solubilized samples of *Blastochloris viridis* photosynthetic reaction center (RC_{vir}). The microjet of solubilized RC_{vir} (green), X-ray detector, XFEL beam (orange) and 800-nm pump laser (red). Reproduced with permission from Arnlund *et al.* (2014).

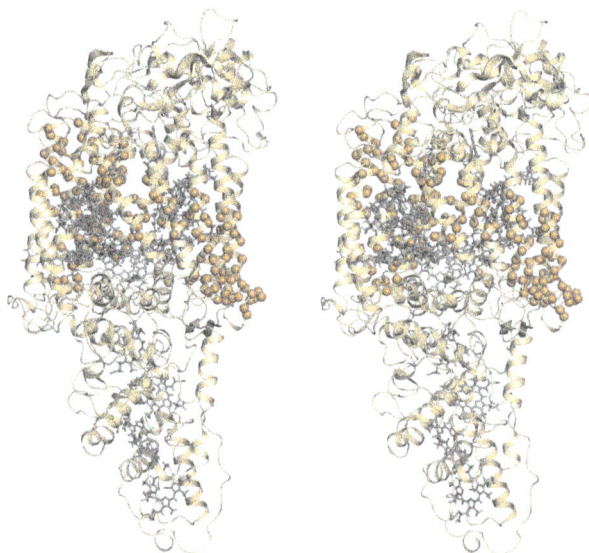

Figure 6.19.   Structural analysis of the protein conformational changes. Stereo representation of light-induced movements in *Blastochloris viridis* photosynthetic reaction center (RC_{vir}). Spheres represent $C\alpha$ atoms that display recurrent movements within an ensemble of structural changes and orange spheres represent $C\alpha$ atoms with recurring movements away from other $C\alpha$ atoms. Cofactors are shown in gray. Reproduced with permission from Arnlund *et al.* (2014).

the 0.2 nm resolution level is shown to be achievable with current free electron laser sources. This accomplishment represents the smallest protein crystals to yield a high-resolution structure by X-ray crystallography. The authors suggest that with these advances in X-ray imaging technology, it should be possible to obtain data from protein crystals with only 100 unit cells using the currently available XFELs. Thus single-molecule imaging of individual biomolecules could almost be within reach!

References

Arnlund, D., L.C. Johansson, C. Wickstrand, A. Barty, G.J. Williams, E. Malmerberg, J. Davidsson, D. Milathianaki, D.P. Deponte, R.L. Shoeman, D. Wang, D. James, G. Katona, S. Westenhoff, T.A. White, A. Aquila, S. Bari, P. Berntsen, M. Bogan, T.B. Van Driel, R.B. Doak, K.S. Kjaer, M. Frank, R. Fromme, I. Grotjohann, R. Henning, M.S. Hunter, R.A. Kirian, I. Kosheleva, C. Kupitz, M. Liang, A.V. Martin, M.M. Nielsen, M. Messerschmidt, M.M. Seibert, J. Sjohamn, F. Stellato, U. Weierstall, N.A. Zatsepin, J.C. Spence, P. Fromme, I. Schlichting, S. Boutet, G. Groenhof, H.N. Chapman and R. Neutze. Visualizing a protein quake with time-resolved X-ray scattering at a free-electron laser. *Nat. Methods* 11: 923–926, 2014.

Bloembergen, N. *Nonlinear Optics*, W.A. Benjamin, Inc., 1965.

Brown, R.A., Y. Yeh, T.S. Burcham and R.E. Feeney. Direct evidence for antifreeze glycoprotein adsorption onto an ice surface. *Biopolymers* 24: 1265–1270, 1985.

Chao, W., B.D. Harteneck, J.A. Liddle, E.H. Anderson and D.T. Attwood. Soft X-ray microscopy at a spatial resolution better than 15 nm. *Nature* 435: 1210–1213, 2005.

Chapman, H.N., P. Fromme, A. Barty, T.A. White, R.A. Kirian, A. Aquila, M.S. Hunter, J. Schulz, D.P. Deponte, U. Weierstall, R.B. Doak, F.R. Maia, A.V. Martin, I. Schlichting, L. Lomb, N. Coppola, R.L. Shoeman, S.W. Epp, R. Hartmann, D. Rolles, A. Rudenko, L. Foucar, N. Kimmel, G. Weidenspointner, P. Holl, M. Liang, M. Barthelmess, C. Caleman, S. Boutet, M.J. Bogan, J. Krzywinski, C. Bostedt, S. Bajt, L. Gumprecht, B. Rudek, B. Erk, C. Schmidt, A. Homke, C. Reich, D. Pietschner, L. Struder, G. Hauser, H. Gorke, J. Ullrich, S. Herrmann, G. Schaller, F. Schopper, H. Soltau, K.U. Kuhnel, M. Messerschmidt, J.D. Bozek, S.P. Hau-Riege, M. Frank, C.Y. Hampton, R.G. Sierra, D. Starodub, G.J. Williams, J. Hajdu, N. Timneanu, M.M. Seibert, J. Andreasson, A. Rocker, O. Jonsson, M. Svenda, S. Stern, K. Nass, R. Andritschke, C.D. Schroter, F. Krasniqi, M. Bott, K.E. Schmidt, X. Wang, I. Grotjohann, J.M. Holton, T.R. Barends, R. Neutze, S. Marchesini, R. Fromme, S. Schorb, D. Rupp, M. Adolph, T. Gorkhover, I. Andersson, H. Hirsemann, G. Potdevin, H. Graafsma, B.

Nilsson and J.C. Spence. Femtosecond X-ray protein nanocrystallography. *Nature* 470: 73–77, 2011.

Chen, X., O. Nadiarynkh, S. Plotnikov and P.J. Campagnola. Second harmonic generation microscopy for quantitative analysis of collagen fibrillar structure. *Nat. Protoc.* 7: 654–669, 2012.

Cheng, J.-X. and X.S. Xie. *Coherent Raman Scattering Microscopy*, CRC Press, 2017.

Cox, G., E. Kable, A. Jones, I. Fraser, F. Manconi and M.D. Gorrell. 3-dimensional imaging of collagen using second harmonic generation. *J. Struct. Biol.* 141: 53–62, 2003.

Da Silva, L.B., J.E. Trebes, R. Balhorn, S. Mrowka, E. Anderson, D.T. Attwood, T.W. Barbee, Jr., J. Brase, M. Corzett, J. Gray, J.A. Koch, C. Lee, D. Kern, R.A. London, B.J. MacGowan, D.L. Matthews and G. Stone. X-ray laser microscopy of rat sperm nuclei. *Science* 258: 269–271, 1992.

Follonier, S., W.J.W. Miller, N.L. Abbott and A. Knoesen. Characterization of the molecular orientation of self-assembled monolayers of alkanethiols on obliquely deposited gold films by using infrared-visible sum-frequency spectroscopy. *Langmuir* 19: 10501–10509, 2003.

Fu, D., G. Holtom, C. Freudiger, X. Zhang and X.S. Xie. Hyperspectral imaging with Stimulated Raman Scattering by chirped femtosecond lasers. *J. Phys. Chem. B* 117: 4634–4640, 2013.

Fu, D., W. Yang and X.S. Xie. Label-free imaging of neurotransmitter acetylcholine at neuromuscular junctions with Stimulated Raman Scattering. *J. Am. Chem. Soc.* 139: 583–586, 2017.

Gati, C., D. Oberthuer, O. Yefanov, R.D. Bunker, F. Stellato, E. Chiu, S.M. Yeh, A. Aquila, S. Basu, R. Bean, K.R. Beyerlein, S. Botha, S. Boutet, D.P. Deponte, R.B. Doak, R. Fromme, L. Galli, I. Grotjohann, D.R. James, C. Kupitz, L. Lomb, M. Messerschmidt, K. Nass, K. Rendek, R.L. Shoeman, D. Wang, U. Weierstall, T.A. White, G.J. Williams, N.A. Zatsepin, P. Fromme, J.C. Spence, K.N. Goldie, J.A. Jehle, P. Metcalf, A. Barty and H.N. Chapman. Atomic structure of granulin determined from native nanocrystalline granulovirus using an X-ray free-electron laser. *Proc. Natl. Acad. Sci. U.S.A.* 114: 2247–2252, 2017.

Ly, S., G. McNerney, S. Fore, J. Chan and T. Huser. Time-gated single photon counting enables separation of CARS microscopy data from multiphoton-excited tissue autofluorescence. *Opt. Express* 15: 16839–16851, 2007.

Schie, I.W., T. Weeks, G.P. McNerney, S. Fore, J.K. Sampson, S. Wachsmann-Hogiu, J.C. Rutledge and T. Huser. Simultaneous forward and epi-CARS microscopy with a single detector by time-correlated single photon counting. *Opt. Express* 16: 2168–2175, 2008.

Scully, M.O., G.W. Kattawar, R.P. Lucht, T. Opatrný, H. Pilloff, A. Rebane, A.V. Sokolov and M.S. Zubairy. FAST CARS: Engineering a laser spectroscopic technique for rapid identification of bacterial spores. *Proc. Natl. Acad. Sci.* 99: 10994–11001, 2002.

Sohrabpour, Z., P.M. Kearns and A.M. Massari. Vibrational sum frequency generation spectroscopy of fullerene at dielectric interfaces. *J. Phys. Chem. C* 120: 1666–1672, 2016.

Stoller, P., K.M. Reiser, P.M. Celliers and A.M. Rubenchik. Polarization-modulated second harmonic generation in collagen. *Biophys. J.* 82: 3330–3342, 2002.

Wei, X., P.B. Miranda, C. Zhang and Y.R. Shen. Sum-frequency spectroscopic studies of ice interfaces. *Phys. Rev. B* 66: 085401, 2002. DOI:10.1103/PhysRevB.66.085401.

Wei, X. and Y.R. Shen. Vibrational spectroscopy of ice interfaces. *Appl. Phys. B: Lasers Opt.* 74: 617–620, 2002.

Williams, R.M., W.R. Zipfel and W.W. Webb. Interpreting second-harmonic generation images of collagen I fibrils. *Biophys. J.* 88: 1377–1386, 2005.

Yariv, A. *Quantum Electronics*, John Wiley & Sons, Inc., New York, 1968.

Chapter 7

Temporal Dynamics

The most direct way to ascertain changes in molecular arrangements using microscopy is to compare one image with another. If the desired feature has shifted in spatial position between the sampling of two successive images, there is clearly a temporal change in the structural alignment of the molecule(s). Quantification of that change is, however, a more difficult problem. When images contain a great amount of spatially distinct information, tracking all of the points of interest continuously in time is a difficult task, if not impossible. In order to selectively monitor certain events or markers in a complex environment, there are several techniques now available. Amongst the many optical methods, there are two that have become the hallmark of techniques for molecular dynamics measurements: (1) particle tracking measurement and (2) time correlation function measurement. In the main, both of these techniques require that the investigator *lock* onto the "particle of interest" by using some type of tag. Thus far, most of the tagging has been done by the use of fluorescence marker, although temporal dynamics done by tracking Raman signatures have been performed. We shall delve into both particle tracking and time correlation spectroscopy here.

7.1 Particle Tracking

Particle tracking is the process of finding a specific, labeled but accurately definable species, and following its time trajectory. It is often noticed that tracked particles "behave" differently in their respectively different environments. Thus information about the nature of the environment, the possible range of interactions this particle has undergone in its traversal, and the ultimate fate of the particle or the tracking label, can all be gained by using this method.

Table 7.1.   Measurable particle motion paths due to different environment parameters (modified from Saxton [2009]).

Normal free diffusion motion in solution	$\langle r^2 \rangle = 2dDt$
Anomalous diffusion motion	$\langle r^2 \rangle = \Gamma t^\alpha$ where $\alpha < 1$
Directed motion	$\langle r^2 \rangle = 2dDt + (vt)^2$
Confined motion	$\langle r^2 \rangle \approx \langle r^2 \rangle_0 (1 - e^{\frac{t}{\tau}})$

We start with a list of theoretical paths that particles with and without a confined and jostling environment might experience. Table 7.1 shows the most common types of motion experienced by "particles" within confined quarters.

Here, given the possible stochastic nature of the particle movement, it is the mean square microscopic displacement given by $\langle r^2 \rangle$ that is most easily measured. The *dimensionality* of the system that the particle resides in is given by d. Thus for three-dimensional free diffusion motion, $d = 3$, hence the mean-square distance for free diffusion is simply $\langle r^2 \rangle = 6Dt$. This "normal diffusion" environment describes one with infinite dilution where the tagged particle encounters only Brownian random collision from the small solute molecules. In "anomalous" diffusion, note that the slowing down of the diffusion motion is given by the exponent α, where $\alpha < 1$.

Often, particles such as cells move toward a source of nutrient in some organized fashion, given by a directed velocity v. In that situation, we note that the directed motion contains both a vt term as well as the random motion term. Finally, if the particle is confined in some manner, then $\langle r^2 \rangle_0$ signifies the confined motion radius, while τ is the time during which that the particle is confined in rather complex paths.

A schematic plot of $\langle r^2 \rangle$ versus time t for the several types of motion is shown in **Figure 7.1**. Here we exhibit the sketches of possible types of motion in two-dimensional space, schematically.

Once plots such as the above sketch of $\langle r^2 \rangle$ versus t have been developed from experimental data, analysis of the specific situation focusing on one of the environments that lead to the measured data can take place. A summary of these are shown in **Figure 7.2** (Saxton, 2007).

Several concerns need to be addressed in monitoring fluorescently labeled particles moving in time. One of the most worrisome points is *photobleaching* that depletes the fluorophore signal of interest. Intense excitation

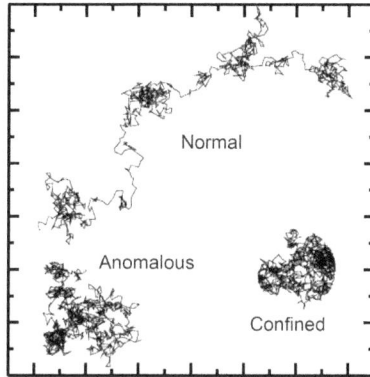

Figure 7.1. Types of motion. Typical trajectories for a random walk of 1,024 time steps for normal diffusion, anomalous subdiffusion with $\alpha = 0.8$, and confined motion in a circular corral. All are to the same scale. Reproduced with permission from Saxton (2009).

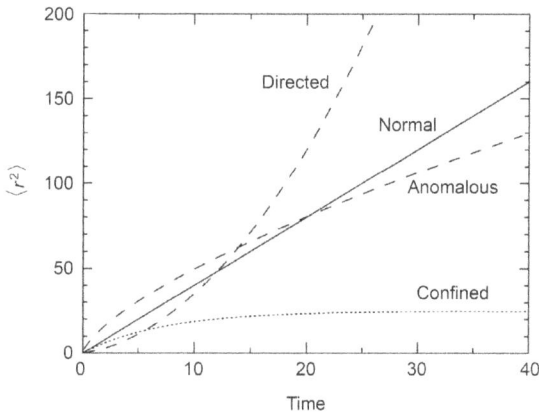

Figure 7.2. Mean-square displacement $\langle r^2 \rangle$ as a function of time for normal diffusion, anomalous subdiffusion, directed motion, and confined motion. Reproduced with permission from Saxton (2009).

of fluorophores very often leads to photobleaching of either temporary or permanent nature. This has to be clearly understood when conducting one of these experiments. On the other hand, at the single molecule level, stepwise photobleaching can be very useful for identifying the source of the fluorescent signal. For single particle tracking, the total disappearance of a moving fluorophore signal may mean the end of some type of activity.

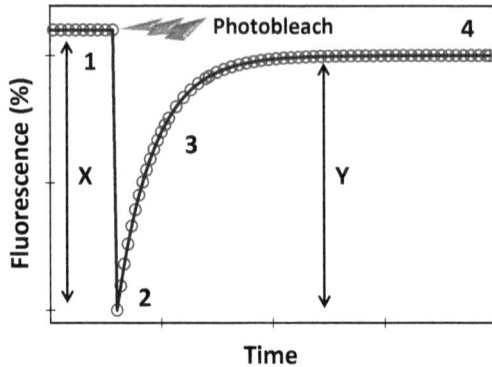

Figure 7.3. Typical fluorescence recovery after photobleaching (FRAP). Adopted from Koppel *et al.* (1976) and Axelrod *et al.* (1976).

On a larger, hence ensemble scale, *fluorescence recovery after photobleaching* (FRAP) has been successfully implemented as a method to monitor molecular dynamics even when these fluorophores do photobleach. In this method, developed by the Webb group (Axelrod *et al.*, 1976; Koppel *et al.*, 1976), it is the recovery of the fluorophore signal derived from the diffusive motion of the non-photobleached particles back into the photobleached domain that provides the quantitation of dynamic process. A simple illustration of the FRAP process is shown in **Figure 7.3**.

In this sketch of the FRAP curve, the X level is the original fluorescence level, and Y represents the asymptotic recovery level. The rate of the recovery is given by the curve starting at time point 2. If the process of fluorophore recovery is dynamic, such as a diffusional process, the slope, by analyzing region 3, will yield the diffusion constant. If the fluorophore is rigidly attached to a molecule of interest, then this recovery provides information about that molecule. Due to the limited depth of focus that is experimentally attainable, most of these studies have taken advantage of the two-dimensional membrane structure to minimize the depth dimension. In that case, the information of diffusional recovery yields lateral diffusion rates.

A second process related to the loss of the fluorophore signal is *resonant energy transfer*. If the excited state of the fluorophore has another non-radiative path for the electron to depopulate, then the fluorescence signal suffers in intensity. The researcher has to be very sure that the depletion of a signal is not due to uncontrolled resonant energy transfer. The transfer may be of the FRET type where the excited fluorophore gets into close proximity with an acceptor fluorophore and exchange via dipole–dipole interaction, as we had discussed in Chapter 3. The signal loss may also be simply the loss of fluorescence because the excited fluorophore has another path for energy transfer without emission. The rapid development of the super-resolution microscopy using controlled blinking pathways of fluorophores that are randomly distributed has helped our understanding of many pathways for the excited electrons to lose its excitation, FRET being just one of them. Many other means exist to cause controlled blinking. We refer to Chapter 5 on the topic of STORM and PALM techniques, where localization with ultra-high precision has been achieved using stochastic photoactivation methods (Sauer, 2013).

Many inorganic quantum dots (QD) have now been fabricated to provide a whole spectrum of emission lines. These are inorganic semiconductors that have been fabricated into very small elements from 2 nm to 10 nm in diameter. Being inorganic molecules in crystalline structures, there is often very little photobleaching. However, because these excited states exist in excitonic levels or conduction bands, there are other pathways by which they can lose the excited electronic energies without recombination radiation (inorganic molecule's fluorescence emission). Such internal energy conversion mechanisms often lead to uncontrolled blinking. Researchers have taken advantage of this blinking phenomenon to be a good hallmark of single fluorophore monitoring. There have also been combination systems such as WS2/MoSe2 bilayer heterostructure that can be constructed to control the blinking process, making these newer constructs useful for fluorescence microscopy methods that require rapid dark-light state conversions (Xu *et al.*, 2017).

Whenever the time sequence of a particular system is tracked, quantified, and analyzed according to these theories that predict dynamics, we gain some information about the environment that led to such detected and measured motion.

7.1.1 *Tracking motor molecule dynamics*

An essential requirement of the particle tag is that in itself it does not interfere with the motion of the particles sampled. In the case of motor proteins, there have been many successful studies tracking the movement of motor molecules. The labeled species, the motor protein itself or cargo on the motor protein, traversing on a track complementary to the motor

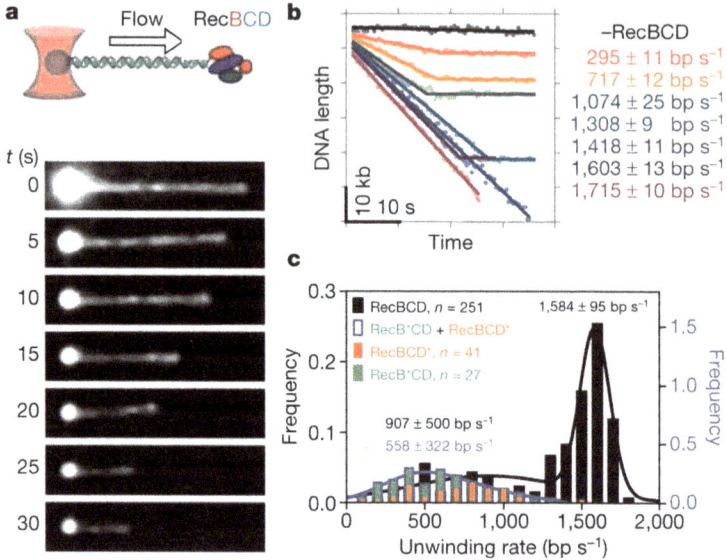

Figure 7.4. RecBCD helicase unwinding of a single dsDNA molecule shown schematically in (a). Time sequence of the unwinding process is shown in the series of timed image frames in (b). Rate of unwinding is shown for different species of RecBCD in (c). Distribution curves of unwinding rates show clear distinction between native and modified versions of RecBCD. Reproduced with permission from Liu *et al.* (2013).

system, e.g., myosin on actin track or kinesin on microtubule track, is followed in time. It could also be the movement of a polymerase on the DNA track, or conversely, the change in a labeled track upon the influence of the motor system, such as the case of the DNA helicase opening up the dsDNA in a processive manner. We had discussed this project in our study of the RecBCD helicase on the opening up of dsDNA as an example in Chapter 4. Using labels that are intercalated fluorophores such as YOYO-1, the opening up of the dsDNA by the RecBCD helicase converts the fluorophore's efficiency downward by 10^{-4} times. Such changes are easily discernable when an established, fluorescently labeled dsDNA track encounters RecBCD helicase.

In **Figure 7.4**, Liu *et al.* (2013) extended the work of Bianco *et al.* (2001) in exhibiting the variability in the rate of helicase activity between wild-type and modified RecBCD motors operating on a dsDNA. In Figure 7.4, this range of variability is clearly shown. The wild-type RecBCD exhibits two Gaussian distributions for unwinding, with the faster rates dominant. The two mutated varieties, one with modification on RecB and the other on RecD, both show predominantly slow helicase activity, suggesting that at least one of the motors in these mutant species is not properly engaging in the motor sense to drive through the opening of the dsDNA. Extensions of this simple method were used recently for determining the degree of resection during DNA repair process (Carrasco *et al.*, 2014).

Such observable "tracks," whether increasing or decreasing in time, are convenient markers for the presence of dynamic processes. It is easy to project this type of particle tracking to other motor systems. At the single molecule level, the tracking of a labeled motor protein, such as myosin or kinesin, may also exhibit time pauses in movement, reflecting the stochastic nature of the molecular activity. An example of the myosin movement on actin is shown in **Figure 7.5** (Warshaw *et al.*, 2005; Yildiz *et al.*, 2003).

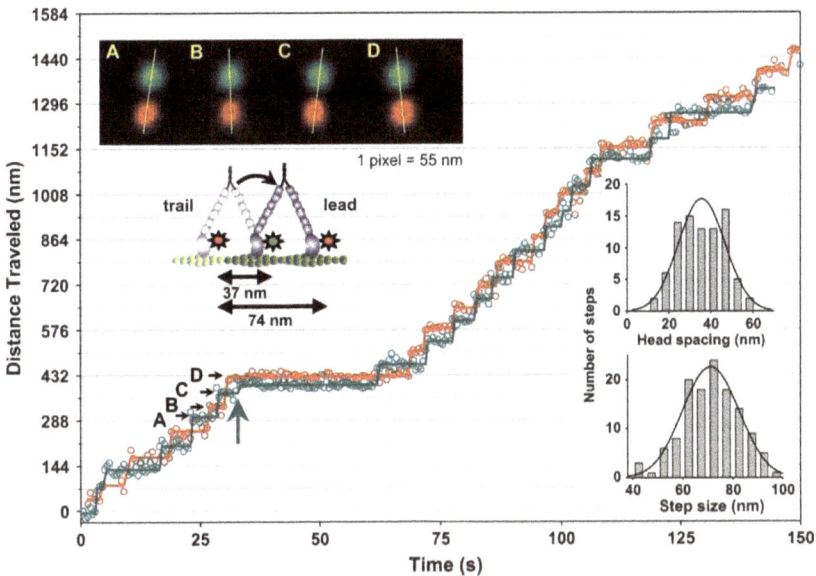

Figure 7.5. Myosin V processive run with heads labeled with different colored quantum dots (Qdots). Green and red open circles are the Qdot565 and Qdot655 positions, respectively, determined by Gaussian fits. Solid lines are the average Qdot positions between steps with the onset of steps determined by eye. On the upper left are averaged Qdot images for steps labeled A–D, with red and green images offset by 12 pixels in y for clarity. The yellow lines connect Qdot centers emphasizing alternating relative head positions. The green arrow identifies substep. On the lower right are histograms of interhead spacing and step size. Reproduced with permission from Warshaw *et al.* (2005).

Here, the tracking of myosin V on actin filament demonstrated that the movements take place in discrete steps. The average size of the steps is 74 nm, which is roughly double the dimension of a single myosin molecule. Such data provide information on the manner by which myosin motors traverse on the actin filament.

In general, particle tracking can be used to monitor the movement of any labeled particles in any condition. Given the utilization of the ultrahigh resolution fluorescence microscopy, the accuracy of these techniques has become very high.

7.2   Time Correlation Analysis

7.2.1   *Photon Correlation Spectroscopy (PCS or Dynamic Light Scattering [DLS])*

In a rigorous interpretation, the PCS method senses anything that produces light or photons and develops a correlation spectrum that is statistically meaningful. Historically, the initial PCS studies were conducted on samples that did not have labels but were only from a light scattering process. Hence the acronym PCS is mostly associated with correlation spectrum from *scattered* species. This is contrasted against the field of Fluorescence Correlation Spectroscopy (FCS), which denotes time correlation analysis of fluorophores. We shall get into FCS later.

For the scattering measurement of temporal dynamics, as mentioned in Chapter 2, the time autocorrelation function is the temporal Fourier transform of the spectral function, via the Wiener–Kinchine theorem. The original studies were performed in the analog spectral mode (Cummins *et al.*, 1964). Principal reviews of the method (PCS or Dynamic Light Scattering [DLS] or Quasi-elastic Light Scattering (QELS)) are contained in three volumes: Chu (1974), Berne and Pecora (1976), and Chu (1991).

Since everything that is even slightly inhomogeneous in the refraction index will scatter light, one has to be very careful to ascertain from what species the detected signal is derived. For this reason, PCS has not been handy for these studies except in the case of particle- or vesicle-size sampling, *in vitro* controlled protein aggregation processes, and if in conjunction with polarization sensitivity, the change in orientation of rigid macromolecules in solution. Polymer flexibility studies have also been performed using PCS, e.g., Fan *et al.* (1987), Fujime *et al.* (1987), and Kubota *et al.* (1987). In each of these examples mentioned, it is the label-free dynamic information related to the rate change of particle diffusion, orientation, or flexure respectively that changes the PCS signal. From such a piece of experimental data, the structure of the molecule (usually in solution) is inferred or deduced.

The theory behind PCS is a comparison of captured photonic (scattered) information sampled at two different times by the use of the statistical probability in time correlation. Suppose that the incident light is being scattered by a sample over the entirety of the illuminated region, and the scattered light is basically at the same optical frequency as the incident. In such a case, we have elastic scattering. This means that incident light and scattered light have the same frequency at the same time. Scattering from sampled point r_0 at t_0 produces a scattered field $E_{sc}(r_0, t_0)$. If now we sample the same point in space, r_0, at a later time, t_1, the scattered field will be $E_{sc}(r_0, t_1)$. Even though the scattered field is still at the same wavelength (elastic), it has picked up a phase change due to movement. Then by calculating the *first order time correlation function*, we get the equation below

$$C(r_0, \tau) = \langle E(r_0, t_0) \cdot E(r_0, t_0 + \tau) \rangle. \tag{7.1}$$

Here τ is a variable covering all the relevant time differentials sampled in the experiment. How do these phase changes occur? If the particle being sampled is being pushed along a flow direction at a velocity $\mathbf{v}$, it will pick up a phase change consistent with $\mathbf{k} \cdot \mathbf{v}$, where $k = 2\pi/\lambda$. A correlation measurement thus will yield the velocity of the particle movement. This conceptual basis is the principle behind *laser Doppler anemometry* (LDA) (Yeh and Cummins, 1964), now broadly used in Doppler radar or Doppler imaging.

The detection of optical radiation, however, is not an **E**-field measurement but an intensity or number-of-photons detection, as all of our known detectors are square-law detectors. Photoemission requires that the photons exciting a photosensitive surface eject free electrons that can then be counted. Thus, the quantity that is directly measured is not the field correlation, but the intensity or numbers correlation for the study of temporal dynamics. Indeed, in order to obtain velocity information from PCS experiments, an *optical heterodyning* configuration has to be used so that even using the square-law photodetectors, there is one component of the measured signal that has the optical phase information. It is this change in phase that is measured when Doppler movement is present.

By designing an experiment whereby both the scattered light and an unscattered light signal is received by the photodetector simultaneously, we have

$$E_{\text{Tot}} = E_{\text{sc}} + E_{\text{ref}} = E_{\text{sco}} e^{i(\omega_0 + \omega_{\text{sc}})t} + E_{\text{lo}} e^{i\omega_0 t}. \tag{7.2}$$

Due to square-law detection of the photodetector, we have

$$I(t) = |E_{\text{Tot}}|^2 \tag{7.3}$$

which leads to the result

$$I(t) = \text{Constant} + 2E_{\text{sco}} E_{\text{lo}}^* \cos \omega_{\text{sc}} t. \tag{7.4}$$

Thus the frequency of the measured signal will reflect the directed velocity of the movement of the scatterer, $\omega = \mathbf{k} \cdot \mathbf{v}$, where $k = 2\pi/\lambda$, as usual. Successful application of LDA in blood flow measurements was first reported by Benedek's group (Riva *et al.*, 1972).

Another important point to consider is that since optical radiation wavelength is in the ~ 500 nm regime, its basic, focused sampling domain is the Airy disc and Rayleigh range along the axial direction. Thus we are typically sampling a large number of molecules. Thus, the $\langle \cdots \rangle$ symbol must represent the ensemble averaged signal from a large number of molecules each of which is scattering light. Finally, because of the square-law detection we just mentioned with respect to LDA, in PCS, we also need to consider the time correlation approach that will yield information through *intensity* correlation function, not field correlation function.

Under such a controlled environment, the intensity or second order correlation function becomes

$$C_2(r_0, \tau) = \langle I(r_0, t_0) \cdot I(r_0, t_0 + \tau) \rangle = \langle n(r_0, t_0) \cdot n(r_0, t_0 + \tau) \rangle \tag{7.5}$$

where the proportionality between intensity and number of photons is made explicit. In measurement devices, a normalization is carried out in this signal, rendering $C_2(r_0, \tau)$ into the normalized second order temporal correlation function

$$g_2(\mathbf{q}, \tau) = \frac{C_2(\mathbf{q}, \tau)}{\langle I(t_0) \cdot I(t_0) \rangle} = \frac{\langle I(\mathbf{q}, t_0) \cdot I(\mathbf{q}, t_0 + \tau) \rangle}{\langle I(\mathbf{q}, t_0) \cdot I(\mathbf{q}, t_0) \rangle}. \tag{7.6}$$

Here, $|\mathbf{q}| = 2k \sin(\theta_{\text{sc}}/2)$ defines the direction that scattering was obtained. Note also that when sampling an ensemble of scatterers, it is the angle of the collection that defines the specific phase shift, not necessarily the specific particle scattering.

If the sampled set of particles (molecules, particulates, etc.) is simply executing random motion and the *principle of ergodicity* remains valid, then there is a relationship between the normalized first and second order temporal correlation function

$$g_2(\mathbf{q}, \tau) = 1 + |g_1(\mathbf{q}, \tau)|^2 \qquad (7.7)$$

where $g_1(\mathbf{q}, \tau)$ is the normalized *first-order* temporal correlation function. What one notices here is that in intensity autocorrelation measurement, the phase change from the first order correlation function is now no longer discernable.

A typical sample of the monodispersed sample will yield a PCS correlation spectrum as in **Figure 7.6**.

In the example, the time scale is plotted linearly to reveal the expected strong τ dependence. The calculated diffusion constant can be related to the particle (molecule) size by

$$D = \frac{kT}{6\pi\eta R_h} . \qquad (7.8)$$

Here, T is the solution temperature in Kelvin, η is the viscosity of the aqueous medium containing the dilute solution of the labeled sample, R_h is the hydrodynamic radius of the fluorescent species.

Because of the need to develop a long-time signal for true baseline measurement and subtraction, the PCS measurement is most often presented in a logarithmic time scale, whereby many orders of time can be seen in the plot. In this manner, the baseline is unquestionably the location where there is no correlation signal. A representative curve may look like **Figure 7.7**.

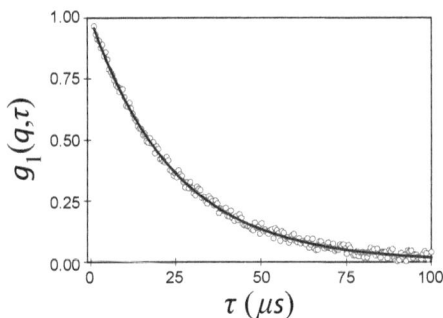

Figure 7.6.   Plot of a $g_1(\mathbf{q}, \tau)$ correlation function for a random diffusion species.

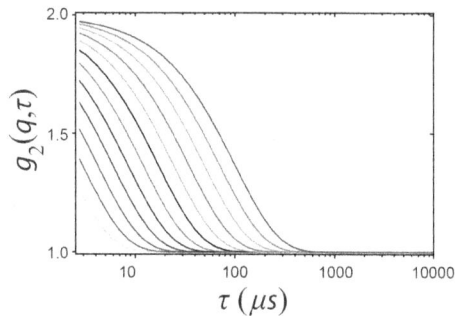

Figure 7.7.   Intensity autocorrelation function. Note that from PCS data using the auto-correlation method, $g_2(\mathbf{q}, \tau)$ lies between the values of 1.0 and 2.0, see Eq. (7.7). The temporal scale is plotted as log of time (τ) to catch the zero-correlation baseline more easily.

In this plot, the $g_2(q, \tau)$ function is plotted against time τ which is in logarithmic scale. The changes in the independent variables: index of refraction variability with a concentration of species change, dn/dc, or time constant, τ_0, have led to the set of multi-colored correlation function plots.

A plot of this nature often does not lend itself to simple analysis unless detailed data fitting has taken place. Many specialized analysis routines exist nowadays from the manufacturer's software package for data analysis and extraction of relevant information from the PCS data files. These include multi-fit analysis, cumulant analysis, and distribution analysis. From these analyses, information about the nature of the scatterers, including size, shape, and even reactivity can be extracted. The main limitation is the fact that sampling requires a relatively large volume, rendering this measurement more of an ensemble one, not close to a single molecule one.

Two ways exist to increase the specificity of PCS measurements. First of all, it is possible to conduct these measurements in a reflective mode instead of transmissive mode. The basis of this is the Michelson interferometer. By conducting PCS in a reflective mode from both a scattered beam and one that is simply reflected but not scattered, we have an interferometric measurement of the signal from scattering. This is the basis of *optical coherence tomography*, which we shall consider in the medical application chapter (Chapter 8).

The other way to increase specificity is to introduce labels. The process is *Fluorescence Correlation Spectroscopy*.

7.2.2   *Fluorescence Correlation Spectroscopy (FCS)*

Fluorescence Correlation Spectroscopy had its direct birth from the ideas of photon correlation spectroscopy. Looking for ways to improve particle specificity that is limiting in PCS, an elegant way to produce photons that is more specific to the species of interest is to elicit fluorescence emission from that sample. Magde, Elson, and Webb (Magde *et al.*, 1972) achieved that by providing labels on certain molecular species in solution and conducting a PCS experiment limiting the detection to only the fluorescent species. Since each molecule is labeled with the same fluorophore, at the same structural location, the translational movement signal strongly represents the movement of the molecule. The measured quantity is the *fluctuation in the fluorescent intensity* $\delta\mathbf{I}$, but this also reflects the number of fluorescently labeled molecules moving in and out of an illuminated domain.

$$C_2(\tau) = \langle \delta I(t_0) \cdot \delta I(t_0 + \tau) \rangle. \tag{7.9}$$

In this diagram (**Figure 7.8**), the incident laser excitation source is focused by an objective of the microscope. Thus the focal volume is again the Airy disk $\times$ Rayleigh range. In (a), the illustration shows the molecules that are being excited within this domain of excitation light volume. Note that certain molecules that are unlabeled (black) will also move in and out of the volume in the same manner that the fluorescently labeled species are doing, but we are sampling only those labeled by appropriate filtering of light. In (b), the measured quantity is presented as the fluctuations in the fluorescent intensity. The autocorrelation curve obtained from such a trace is given in (c). Here, the $C_2(\tau)$ correlation profile for that focused domain is shown. For purely translational diffusion motion, the theoretical expression for the normalized correlation function is $G(\tau) = C_2(\tau)/I^2$

$$G(\tau) = \frac{1}{N} \left[\frac{1}{1 + \frac{4D\tau}{r^2}} \right] \left[\frac{1}{1 + \frac{4D\tau}{l^2}} \right]^{\frac{1}{2}} \tag{7.10}$$

where N is the number of fluorophores in the sampling volume on the average, r is the radius of the Airy disk, and l is the $\frac{1}{2}$ distance of the axial Rayleigh range. Note that in a fluid, these diffusing particles are assumed to execute isotropic translational diffusion.

A typical FCS apparatus is shown in **Figure 7.9**. In this configuration, a high NA(1.4) objective is used to create a small sampling volume in the solution. Here, the back emission of fluorescence is collected by passing the signal through a dichroic mirror and then focused onto an avalanche photodiode (APD) behind a pinhole. This confocal design provides the advantages of selective sectioning of the sample. The correlation curve is then calculated for all the τ channels in parallel.

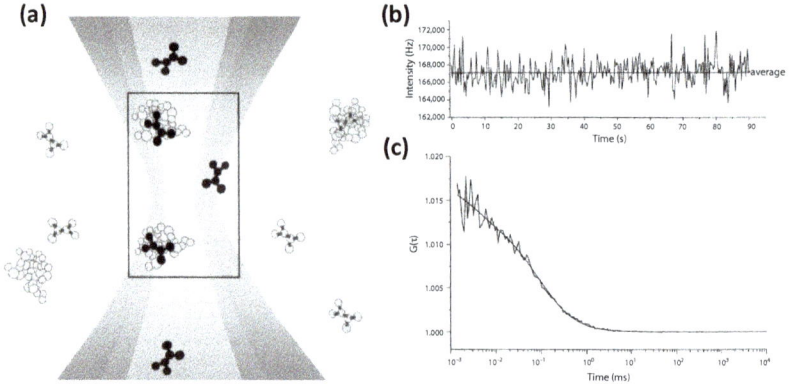

Figure 7.8.   Fluorescence Correlation Spectroscopy. (a) Confocal illumination and detection volumes. Only those particles that diffuse into the confocal volume will be detected by the photon counter. (b) Fluorescence intensity fluctuates over time around an average intensity. (c) Autocorrelation curve shows the decay of temporal correlation over time. Reproduced with permission from Bulseco and Wolf (2007).

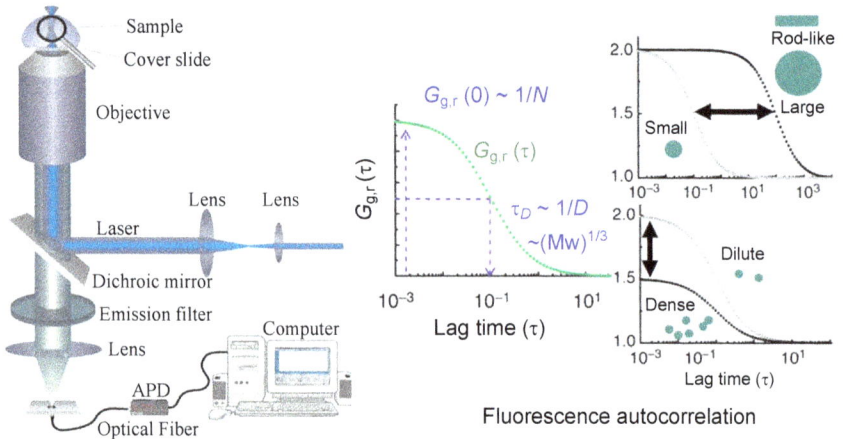

Figure 7.9.   Left: Fluorescence Correlation Spectroscopy (FCS) setup. Reproduced with permission from Schwille and Haustein (2004). Right: Autocorrelation FCS signals, here showing that $G_{g,r}(\tau)$ will provide information on number, density, and size. Reproduced with permission from Pack *et al.* (2014).

Fluorescence Correlation Spectroscopy autocorrelation curves look like that shown on the right side of Figure 7.9. Here, the correlation function is obtained by using either green or red fluorophores, hence, $G_{g,r}(\tau)$. It can be seen clearly that density of the fluorophores is measured on the vertical scale, with low density leading to higher fluctuation signal, consistent with the $1/N$ theoretical derivation. Also shown is the sensitivity of FCS to particle size. The larger ones yield curves with longer time constants τ.

7.2.2.1 *Special concerns of FCS*

(1) Sensitivity to the time dynamics of the fluorophore — Fluorescent probes attached to a molecule may be executing a motion that reflects its own movement or that of the molecule it is designed to probe. A well-designed "probe molecule" is one that will attach to a location that would be innocuous to the functions of the molecule being probed, and does not contain any motion independent of the probed molecule. The idea of thiol-group derived fluorophore probe attachment is one of the most favorable as a probe. The strength of the attachment is high, and the thiol-group is usually not located at the protein's functioning domain. The use of Alexa, ATTO, Cy3,5, Rhomadine, or TR type of dye molecules are all effective when attached to such locations.

(2) Sensitivity to the environment of the biological molecule — In Chapter 2, we have discussed the ways by which the excited state of the electron can return to the ground state that included the direct electric dipole (E1) decay from that state. We mentioned that the other means are non-radiative, involving de-excitation by collision with the solute molecules $k_c = 1/\tau_c$, by transferring from the Singlet manifold to the triplet manifold k_{ST}, or by having another fluorophore nearby to transfer via dipole–dipole interaction k_{FRET}. In total, the non-radiative transition rate $k_{NR} = k_c + k_{ST} + k_{FRET}$. Thus a plot of time evolution in FCS is useful in assessing the presence of these competing modes of fluorescence signal changes in time.

7.3 Progress in the Extension of the FCS Method

For more complex solutions, for example, if more than one species exists, but the labeled species is on species A of a bimolecular reaction of $A + B = AB$. Since the species that can be seen in this single-label FCS is only the labeled species, and we assume that labeling does not hinder the reactivity of the above reaction, the multiple species undergoing reaction may show a compound FCS spectrum described by

$$G(\tau) = \frac{\sum N_i B_i^2 D_i(\tau)}{(\sum N_i B_i)^2} \tag{7.11}$$

where B_i provides the probability that the ith species is properly labeled to be seen by the incident illumination (**Figure 7.10**).

Here, the red curve shows the composite FCS spectrum. At least two decay times are clear from this plot: a slow time ~ 1 ms and a shorter time decay region at nearly 1 μs. The authors have attributed the long-time decay to motional processes and the short-time decay to triplet fraction decay process. Note that amplitudes of the respective contributions are also discernable.

Single-label FCS can also be used to characterize the conformational state of a large molecule. Should a molecular system undergo a reaction from helix to coil conformation, will the hydrodynamic radii change significantly? This may be discernable using the labeled species consistent with the dynamic equilibrium $[helix] \leftrightarrow [coil]$. An example is shown in the following figure where native (globular) shaped molecule converts to a random coil conformation (or *vice versa*).

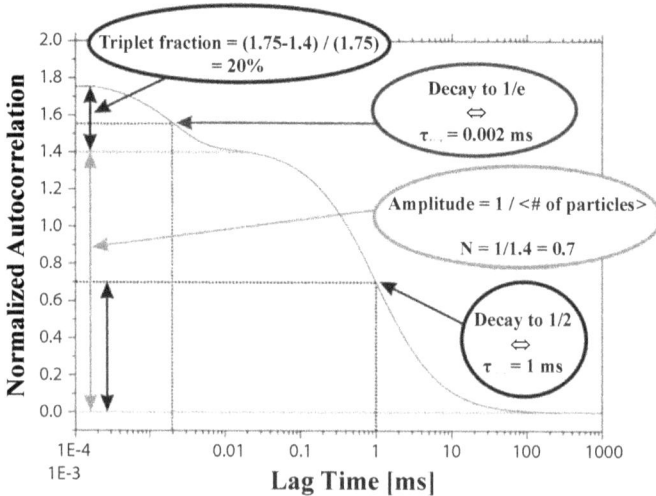

Figure 7.10. Autocorrelation curve with fast triplet dynamics and the role of different parameters. Reproduced with permission from Schwille and Haustein (2004).

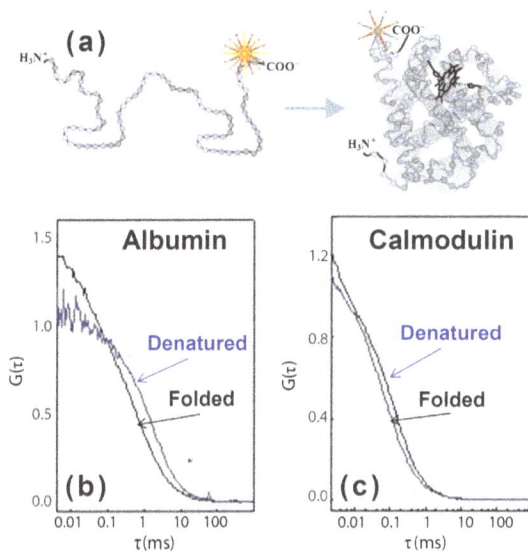

Figure 7.11. Protein folding events by FCS. (a) Proteins are compact in their folded state in comparison with the unfolded (denatured) states. Changes in the diffusion coefficient in the case of (b) albumin and (c) calmodulin upon denaturation. Reproduced with permission from Schwille and Haustein (2004).

From these illustrative plots in **Figure 7.11**, it is clear that the correlation time for the denatured states of both albumin and calmodulin is longer than that of the native states, reflecting a more loose structure of the denatured state, hydrodynamically.

Fluorescence Correlation Spectroscopy can now be carried out within the cell by specifically labeling the components to be sampled. In the example shown in **Figure 7.12**, Schwille's group has shown different mobilities of the different elements within the cell.

Another exciting possibility is to create "reduced cell" configurations and to induce reactivity of proteins in those tightly controlled environments. One such configuration is the *nano-lipidic disc (NLD)* configuration whereby the small membrane construct is stabilized in solution by a scaffolding membrane protein, Apo-lipoprotein E4. In such a configuration, the NLD is stable in solution.

In the example of **Figure 7.13**, suppose one introduces membrane proteins that forms a functioning protein complex on the lipid surfaces of this nanodisc. We have essentially a construct that has a different shape

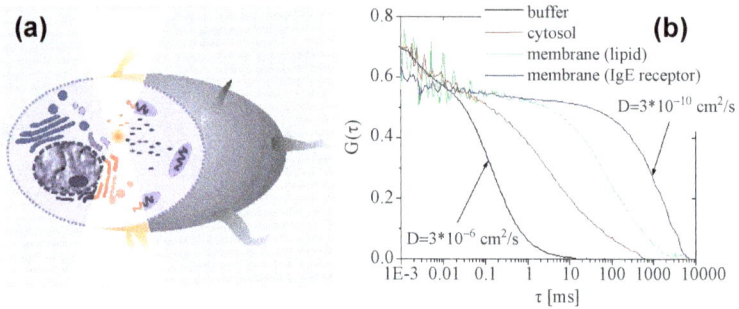

Figure 7.12.   Determination of mobility-related parameters. (a) Fluorescence Correlation Spectroscopy experiments in cellular systems targeting different regions such as cytosol versus membrane. (b) Difference in the measured diffusion constants between the different components. Reproduced with permission from Schwille and Haustein (2004).

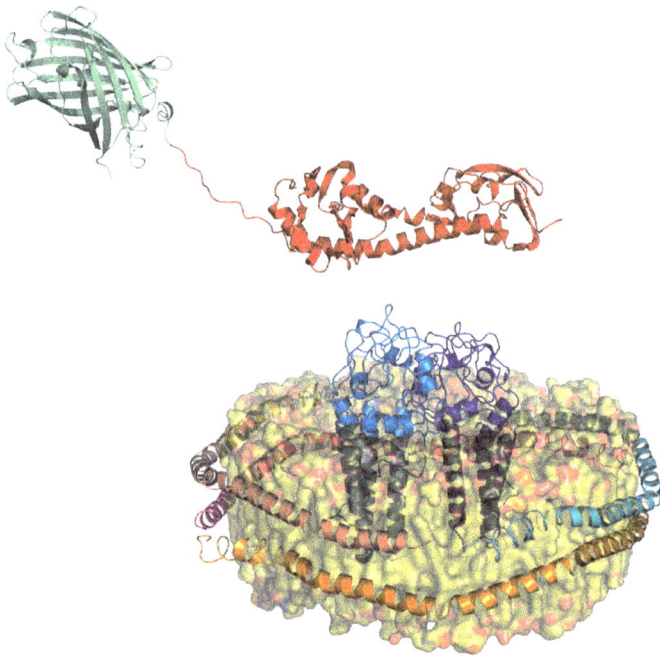

Figure 7.13.   Model of YopB (blue) inserted into a 10-nm nanodisc with cognate protein LcrV (red) labeled with EGFP (green). The molecular masses used in this model are: LcrV (35 kDa), EGFP (27 kDa), YopB (42 kDa, monomer), and nanodisc-YopB complex (258 kDa). The small LcrV binds to the much larger YopB-nanodisc complex, resulting in a significant shift in the autocorrelation curves to longer diffusion times. Reproduced with permission from Ly *et al.* (2014).

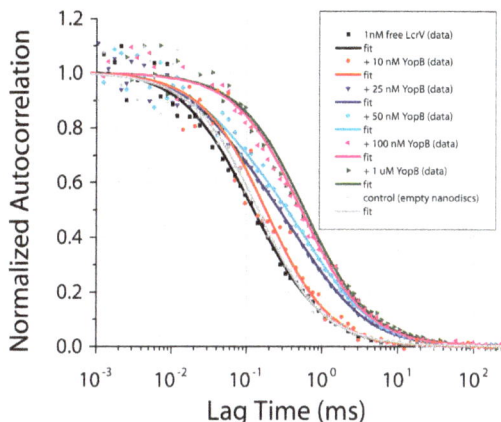

Figure 7.14. Autocorrelation function from FCS. Fluorescence Correlation Spectroscopy data showing the change in correlation time as YopB complex is associated with nanodisc with increasing concentrations. Extremal spectra are for the empty nanodisc and the free protein complex. Reproduced with permission from Ly *et al.* (2014).

compared with the flat nanodisc by itself. These distinct shapes will yield different solution translational diffusion constants. The differentiation of these two constructs forms the initial characterizing of distinctions between a passive and a functioning membrane protein system.

In a study reported by Ly *et al.* (2014) the dynamics of the pure protein complex, the membrane vesicle alone, and the combination upon reaction show very different spectra, as can be seen in **Figure 7.14**, providing a first step toward the goal of studying reactivity in a solution environment.

7.4 Fluorescence Cross-Correlation Spectroscopy (FCCS)

Noting that a binary reaction contains at a minimum, two reacting species, we can also conduct an FCS experiment that would measure the cross-reactivity of these two species. In such a situation, we can label A with red (R) and label B with green (G). Before the species are combined, these species will diffuse independently in a solution, essentially uncorrelated. However, once combined, the red and green signals will exhibit a strong correlation with each other.

If one carries out this fluorescence *cross-correlation* experiment,

$$C_c(\tau) = \langle \delta F_R(t_0) \cdot \delta F_G(t_0 + \tau) \rangle \tag{7.12}$$

and this signal is less than unity, we have measured the level and rate of chemical dissociation reactivity using the cross-correlation measurement. In such a measurement, whenever the species have combined into AB, the binary product, there should be a perfect correlation for all times $(C_c(\tau) = 1)$. Any signal departure from unity is an indication of other processes rendering the two fluorophores not time correlated. This may be a measure of reactivity rate.

A typical cross-correlation apparatus is shown in **Figure 7.15**. Here, the sample to be interrogated has the compound labeled species. Excitation of species 1 is by laser 1 and excitation of species 2 is by laser 2, and one takes care not to have cross-excitation in the design of the experiment.

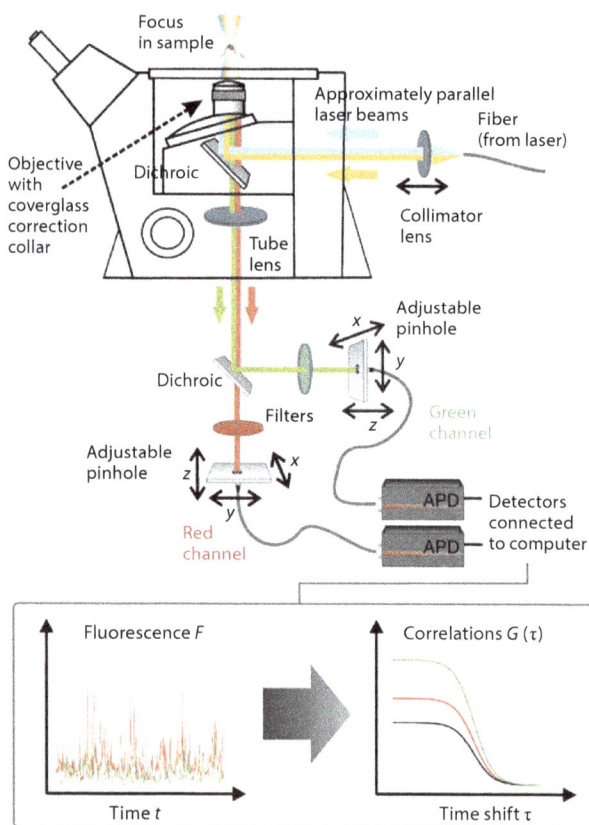

Figure 7.15.   Cross-correlation setup with dual colors. Reproduced with permission from Bacia and Schwille (2007).

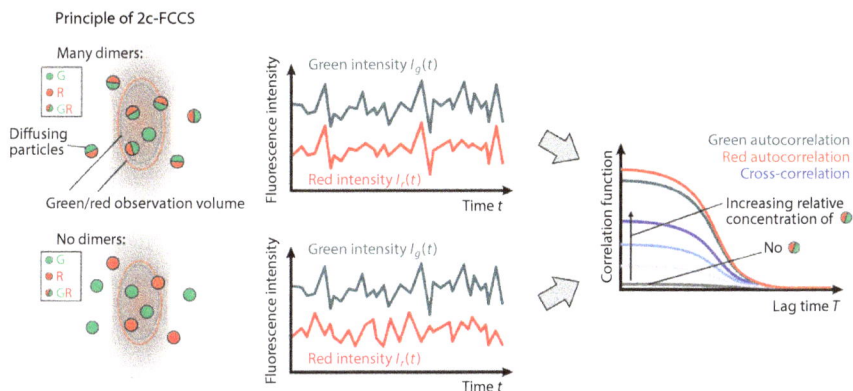

Figure 7.16.   Principle of FCCS dual colors. Reproduced with permission from Krieger *et al.* (2015).

In the above example, the level of cross-correlation is shown by the changing spectrum, noting that any increase in this spectrum signifies cross-reactivity. Thus this measurement is indeed a measure of reactivity between the two species (**Figure 7.16**).

In studies by Schwille's group (Schwille and Haustein, 2004), they have been able to conduct cross-correlation studies within a living cell, thus providing information about the molecular relationships with labeled species within a cell cytoplasm. We mention for reference a recent article by Pack *et al.* using FCCS to measure the cytoplasmic assembly of the 26S proteasome in a live cell (Pack *et al.*, 2014).

7.5   Combining Techniques

7.5.1   *Multi (or Two) Photon Excited Fluorescence (M(T)PEF) FCS*

One additional experimental design improvement has been the use of *two-photon excited fluorescence (TPEF)*. As we had discussed in the earlier chapter (Chapter 6), multiphoton excitation has a similar effect as that of confocal apertures in that it allows more localized excitation of the species, and in a cell, this is an essential criterion. The ability to traverse the cross-section as well as the depth of a cell interior is of paramount importance for intracellular species localization. Thus intracellular temporal dynamics studies have been accomplished using simultaneously confocal microscopy and M(T)PEF.

7.5.2   *Evanescent wave-FCS*

Fluorescence Correlation Spectroscopy in solution, even by using the best of microscope objectives, still covers a rather large spatial domain, and since the sampling is only for species that are labeled, it is highly desirable to reduce the volume of sampling. Two obvious ways that this is done is to use (1) totally internal reflection (TIR) evanescent region sampling or (2) zero-mode waveguide sampling. We have discussed the excitation of a solution system using evanescent wave coupling for TIR excitation in Chapter 2. It is an easy extrapolation to allow the species coming upon an anchored receptor on a surface to be fluorescently labeled, and hence TIR-fluorescence (TIRF) microscopy studies can be conducted (**Figure 7.17**) on this species. In the objective-based TIRF microscope, an annular ring of excitation light incident at larger than the critical angle is used to excite

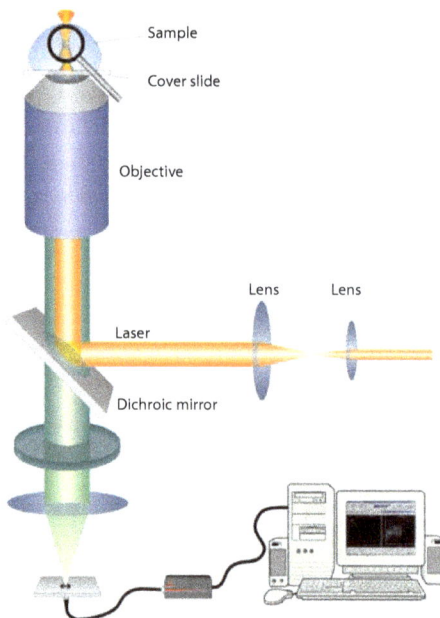

Figure 7.17.   Two-photon TIR setup with excitation from the outer green ring and the collected light at the center. Reproduced with permission from Schwille and Haustein (2004).

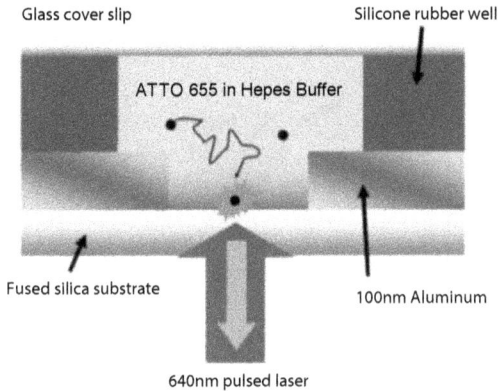

Figure 7.18. Evanescent coupling of fluorescently labeled species. Image obtained through personal communication with Y. Yuen.

the evanescent waves. Backscattered fluorescence from particles within the evanescent region is then collected for FCS spectral analysis. Because of the TIR localization criterion, ~ 100 nm, background non-reacting species will not be visible. This method is thus useful for probing receptor-antigen reactivity in an almost background-free environment.

In the extreme form of this scheme, one imagines the surface to be covered by a set of microscopic apertures, each of which is of dimension much less than the wavelength of visible light (~ 20–50 nm). These nano-structures will not allow free propagation of light waves at all, even with the normal incident light. However, evanescent waves will nonetheless exist within the nano-apertures (Levene *et al.*, 2003). This type of zero-mode waveguide (ZMW) aperture for evanescent wave imaging is shown in **Figure 7.18** (Fore *et al.*, 2007).

In this scheme, because the ZMW opening is so small (< 100 nm), no propagating wave in the visible range can pass. Hence, only evanescent waves exist within the small cavity. Molecules with fluorescent tags that are captured near the bottom of the well within this cavity can couple that evanescent wave and produce fluorescence signals. Note that the incident wave (here at 640 nm) no longer needs to satisfy the critical TIR angle condition.

In **Figure 7.19**, a membrane was deposited within a zero-mode waveguide well. Evanescent excitation of labeled membrane groups show that their FCS signals reflect distinctions in binding affinity. Note that mem-

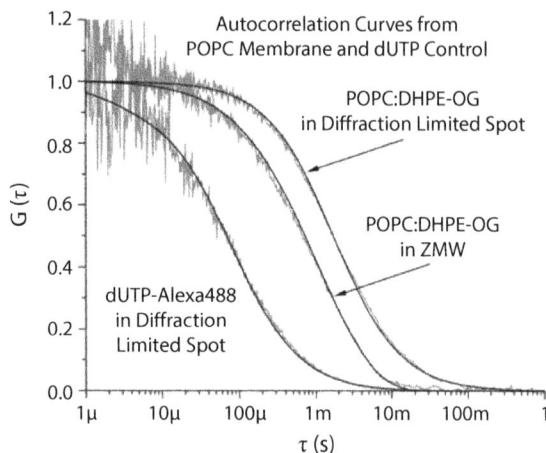

Figure 7.19. Normalized autocorrelation curves from POPC/DHPE-OG and dUTP-Alexa488 in bulk and POPC/DHPE-OG in a ZMW. Curves are shaded with fits shown in black. Left to right: dUTP-Alexa in a diffraction-limited volume, POPC/DHPE-OG in a zero-mode waveguide and POPC/DHPE-OG in a diffraction-limited volume. A three-dimensional FCS model including a triplet component was used for the dUTP curve. The POPC curves were fit using the ZMW model or a two-dimensional FCS model as appropriate. Reproduced with permission from Samiee *et al.* (2006).

brane signals within the ZMW differ considerably from the signal of the same membrane on other supported surfaces. Thus these ZMW structures are becoming tools for highly localized sampling.

One application using these ZMW apertures is to sample differing reactivities within different cells. This idea is the basis of a scheme that Korlach's group is using to conduct DNA sequencing studies using single-molecule fluorescence scheme (**Figure 7.20**).

In this approach, Figure 7.20A provides the basic DNA *template* design. This is a mini-circle DNA template contained in a single guanine site, allowing incorporation of a base-linked fluorescent nucleotide, Alexa Fluor 488-dCTP. In Figure 7.20B, ZMW nanostructures were first treated with polyvinylphosphoric acid (PVPA), enabling selective immobilization of DNA polymerase at the bottom of ZMWs, thus allowing DNA extension reactions. The ZMW observation volume is against the quartz (SiO_2) surface. In Figure 7.20C fluorescent DNA products were imaged from both sides of ZMW arrays. The length of the DNA can be determined by the single-molecule fluorescence brightness analysis. Thus the DNA sequencing process can be monitored.

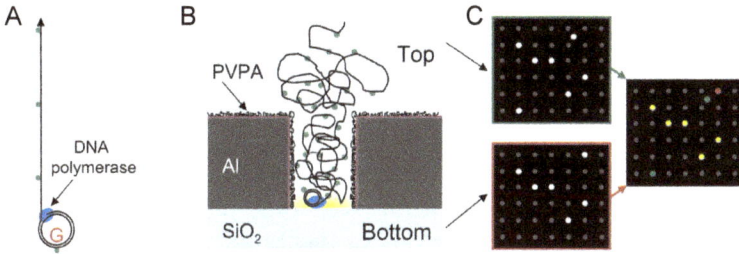

Figure 7.20.   Principle of observing DNA synthesis inside ZMWs. Reprinted with permission from Korlach *et al.* (2008).

7.6   Anti-bunching Spectroscopy

In discussing Fluorescence Correlation Spectroscopy, we are always dealing with the emission of the excited species, which is the labeled species. In an extreme consideration where the sampled domain contains only a single fluorophore, we actually encounter a situation of numbers correlation that is on the extreme side. In conducting FCS experiment of the single molecule, imagine correlation sampling of that single fluorophore in times shorter than the emission lifetime of the fluorophore. In that situation, the single molecule has been excited but has yet to have the time to emit, but if we sample in that time regime, usually $< \sim 1$ ns, there is no signal. Hence the time correlation function drops to zero! This is the so-called "anti-bunching" domain.

Sampling in this temporal regime will lead to additional information in the following manner: consider first, the expression for FCS in this short time span. Without the fluorophore executing any motion-related dynamics (translational, rotational), we have, for continuous excitation,

$$G_2(t) = \frac{N(N-1)}{N^2} + \frac{1}{N}(1 - e^{-\frac{t}{\tau}}). \tag{7.13}$$

Here, N is the number of fluorophores in this system and τ is the lifetime of the excited state of the fluorophore. The schematic spectrum will show the sharp dip at the region of the anti-bunching time regime (**Figure 7.21(a)**). Indeed, this profile exhibits the lifetime of the excited single fluorophore. Now if the fluorophore is being excited by a pulsed laser source, with repetition rate longer than the lifetime of the fluorophore, the anti-bunching signal will be described by Figure 7.21(b):

$$\frac{N_c}{N_l} = 1 - \frac{1}{N} \qquad (7.14)$$

where the correlation signal yields a simple relation showing only the number of fluorophores being sampled. Here in Eq. (7.14), when $N = 1$, the correlated signal is 0. The plot of such a FCS trace will yield the missing spike pulse at time 0. Note that all other time points will show the full correlation. This particular result allows one to use this anti-bunching correlation profile to be a method for determining *molecular stoichiometry* in basic molecular reactions.

If we let $N = 2$, Eq. (7.14) will yield $\frac{1}{2}$ or 0.5 for the time 0 anti-bunching value. Extending to $N = 3$, Eq. (7.14) gives $\frac{2}{3}$ or 0.67. Thus the zero-time point correlation value is an indication of the degree of assembly of the fluorescent species.

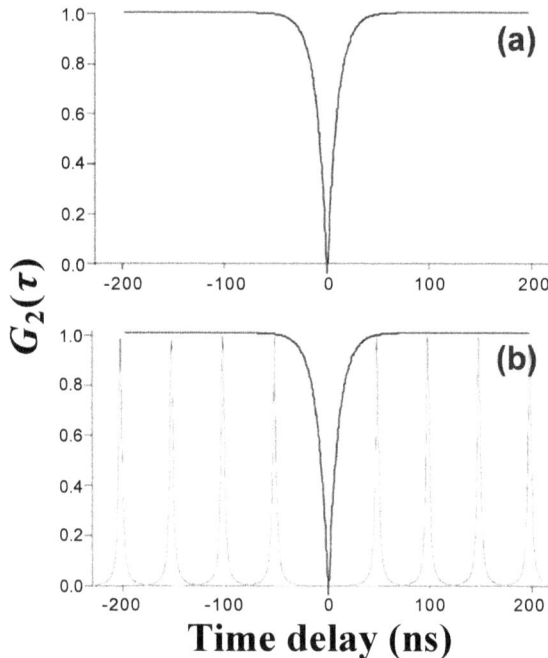

Figure 7.21. Calculated photon pair arrival time histogram for a single fluorescent molecule. (a) Continuous excitation. Note the dip at $t = 0$ indicating anti-bunching. (b) Excitation at 20 MHz pulse repetition (50 nsec) rate. This results in peaks in the photon pair arrival time histogram. Here the anti-bunching signal at time $t = 0$ is shown as 0. Reproduced with permission from Ly *et al.* (2011).

Fore *et al.* (2006) conducted such an experiment on monodispersed hairpin DNA samples labeled either with a single fluorophore, two, or three fluorophores. This sample is deposited on a glass substrate and the antibunching FCS experiment is carried out. In **Figure 7.22(a)**, the image of a typical sampled region is shown, indicating individual hairpin DNA locations on the surface, each with either 1, 2, or 3 Atto-655 dye molecules attached. Sampling each of the bright spots yielded the subsequent set of correlation profiles. In (b), we note that there is a clear 0 signal at the time $t = 0$ point, indicating that the particular point sampled contained a single fluorophore. In (c), the $t = 0$ time location showed a nonzero signal suggesting that more than one fluorophore existed in that bright spot. Statistically, this point is shown to exhibit $N_{ab}/NL = 0.58$, which is somewhat higher than the 0.5 theoretical value for a spot containing two fluorophores. In (d),

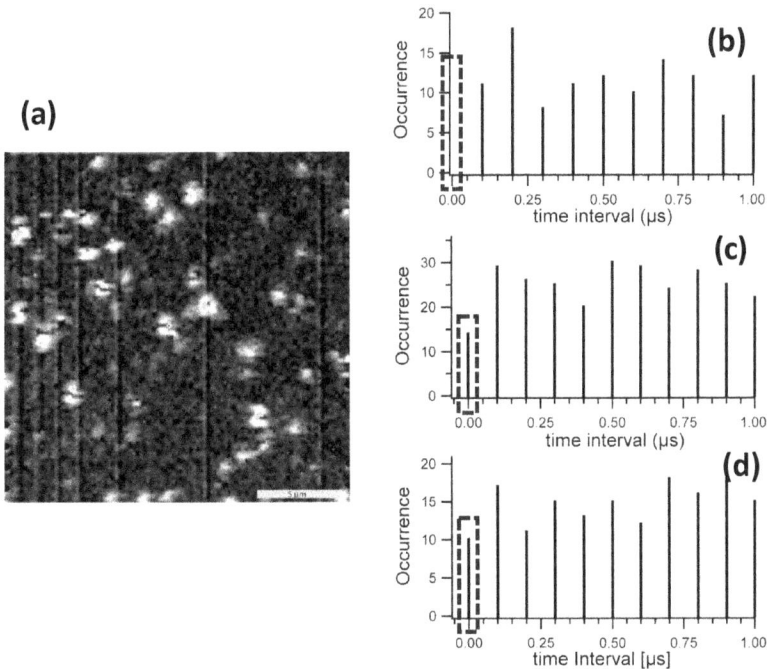

Figure 7.22. (a) 20 μm $\times$ 20 μm image of a DNA hairpin sample, where each DNA hairpin was labeled with just one Atto-655 dye. Scale bar represents 5 μm. (b)–(d) Sample correlation plots for individual spots in an image file similar to the one in image (a). Occurrence = N_{ab}/NL. N_{ab} = Number at 0 lag time representing anti-bunching (dotted boxes); NL = Labeled molecules. Fore (unpublished thesis, 2006).

the $t = 0$ signal showed $N_{ab}/NL = 0.70$, nearly that of the theoretical 0.67 for three fluorophores.

Extending this method for identifying structural stoichiometry, anti-bunching correlation spectroscopywas carried out to determine the number of Apo-A1 proteins wrapping the nanodiscs of high density lipoprotein HDL nanoparticle. Labeling the Apo-A1 with Alexa-647, and waiting for spontaneous equilibrium to be achieved between the DMPC lipid bilayer and Apo-A1 lipoproteins, the results obtained by Ly *et al.* (2011) show that anti-bunching spectroscopy revealed $N = 2$ in these HDL constructs reliably (**Figure 7.23**). In Figure 7.23(a), a controlled single-labeled sample revealed almost zero correlation signal at $t = 0$. In Figure 7.23(b), the $t = 0$ correlation signal is about $\frac{1}{2}$, as one expects for having two apo-A1 associated with one HDL nanoparticle.

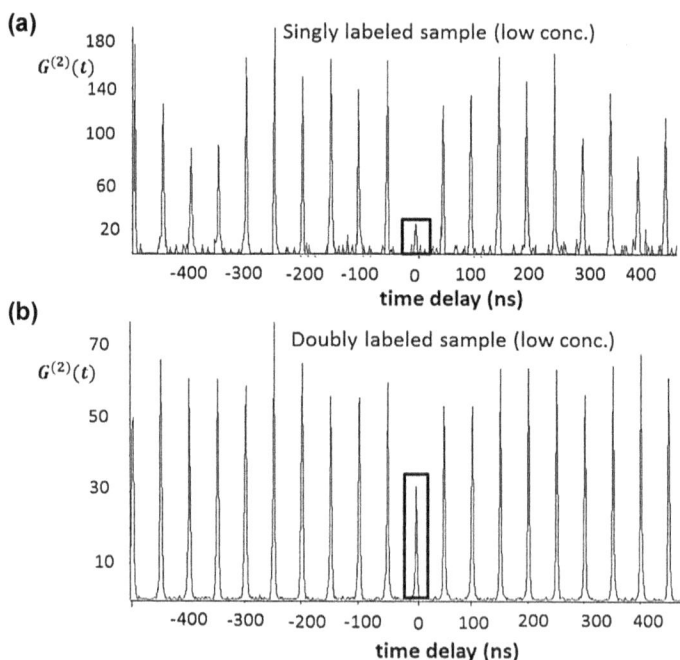

Figure 7.23.   Photon anti-bunching histograms of lipid-bound apoA-I at low concentration. (a) The singly labeled (control) sample shows a very small central peak due to single-molecule emission. (b) The doubly labeled sample (of interest) has a central peak with approximately half the area of the lateral peaks, which is indicative of two protein molecules being present in the sample at any time. Reproduced with permission from Ly *et al.* (2011).

Care has to be given using this approach when the labels are closer to each other than that of the large HDL particle case. This has been shown by Fore *et al.* (2005) when they labeled DNA hairpin structures with Atto dye (655 nm) molecules. They demonstrated that singly labeled guanine in a DNA hairpin configuration could be differentiated from doubly labeled species or triply labeled species. However, for more fluorophores, the theoretical result is not as accurate in this experiment.

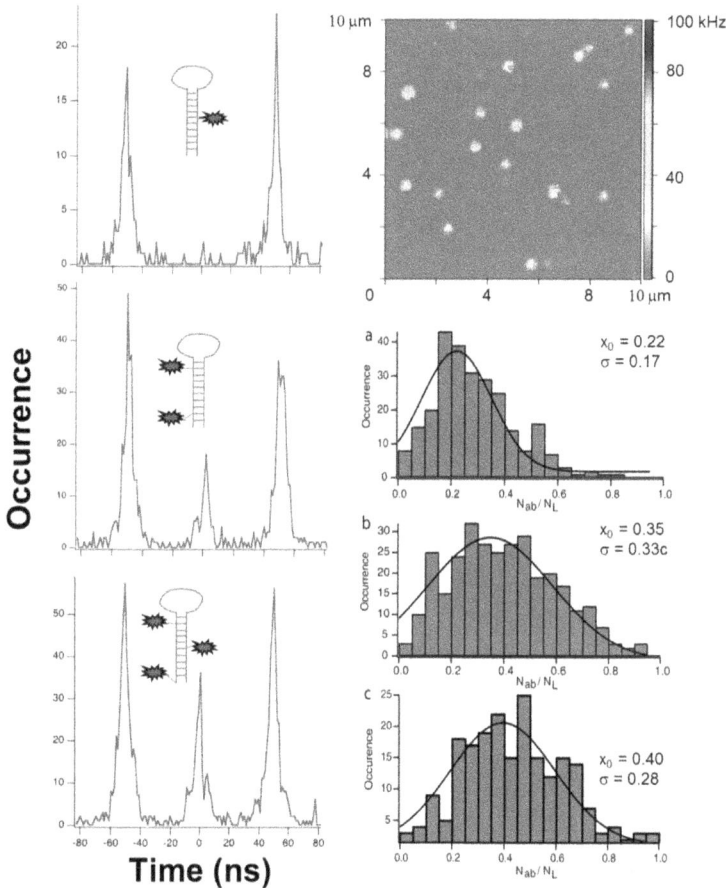

Figure 7.24. Hairpin DNA attached by 1-, 2-, 3-Atto-655 dye molecules. Anti-bunching (AB) time correlation shows the zero-time signal compared to the averaged long-time signal (> 50 ns). Distribution of this AB signal for sampling of larger set (see image inset) is shown in graph, where a,b,c panels reflect the distribution analysis: $N = 1 \rightarrow 0.22$; $N = 2 \rightarrow 0.35$; $N = 3 \rightarrow 0.40$ (Fore, 2006).

In **Figure 7.24**, the DNA hairpin constructs have been illustrated on the left side. The size of this hairpin is significantly smaller than the HDL particle. Even though care has been taken to count only the single spots on the micrograph (background in red), the statistical sampling over a large number of spots yielded the distribution plots on the right side. In particular, we note that for single Atto-655 dye labels, the distribution is not 0.0, but actually yielded 0.22. Correspondingly, for two dye molecules, the expected value is 0.5, but the results show 0.35. For three dye molecules, an experimental value of 0.40 was obtained, not the theoretical 0.67. One principal complication is the possibility that the multiple fluorophores are interacting with each other by being so close in proximity via the self-resonant signaling process. Another is the complication caused by the photobleaching process, which must be more thoroughly examined in a case-by-case situation.

7.7   Imaging FCS

The idea of examining either one image or a serial time sequence of many images and getting meaningful quantitative data on molecular dynamics is the basis of the method called Image FCS (IFCS) (Wiseman and Petersen, 1999). In this situation, the sampling is not a single point, but all of the high-resolution data entries of a CCD image is analyzed. The analysis behind this method yields an expression

$$C(x', y', \tau) = \langle i(x, y, t) | i'(x + x', y + y', t + \tau) \rangle. \qquad (7.15)$$

In this expression, reference with respect to space/time point (x, y, t) and all other data entries are correlated. Thus the space-time correlation entries at $(x+x', y+y', t+\tau)$ for all (x', y', τ) relative to the point (x, y, t) are calculated and stored. Whenever one is seeking spatial image correlation, one relies on the information from a single CCD entry set, whereas when we are looking for the time evolution of a single point, we can track the movement in space and time (successive images).

We consider for example, first of all, spatial correlation at a single time. In that case, we seek the *cluster density* (cd) of the species that is labeled i. This is written as

$$cd_i = \frac{\bar{N}_p}{\pi w^2} = \frac{1}{g_i(0, 0)\pi w^2} \qquad (7.16)$$

where w is the sampling domain radius. We now take a look at this function in its image and correlation plot in **Figure 7.25**. For the image of a healthy retinal pigment epithelial (RPE) cell shown in Figure 7.25(a), the cluster density analysis plot is shown in Figure 7.25(c), showing that there is only a small degree of spatially correlated events, consistent with a random set of fluorophores on the surface of the membrane.

In the next set, Figure 7.25(b) shows that when the RPE cell was induced to initiate cellular apoptosis, a cluster of lipid raft shows up at one pole. In this image presentation, clustering of lipid-rich domains that contained the fluorophores is visually evident. This clustering leads to the spatial IFCS correlation plot showing a single peak of correlation degree ~ 0.35, as shown in Figure 7.25(d). Such analysis provides additional quantitative information about images related to significant biological features being examined.

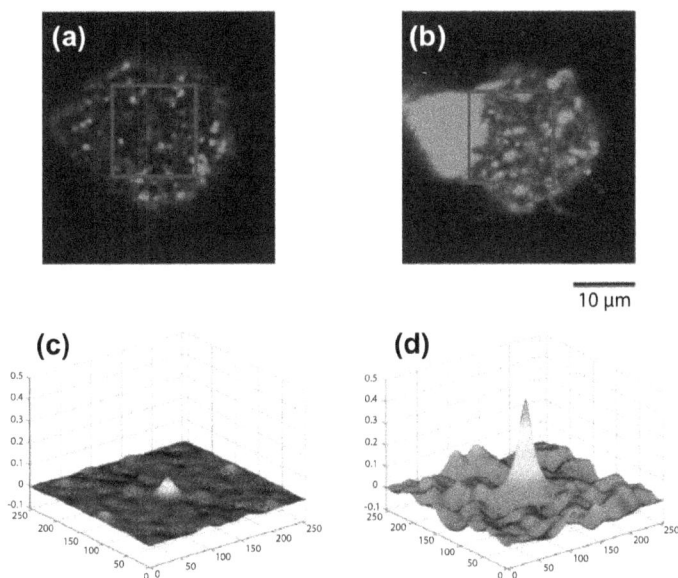

Figure 7.25. (a) Native, healthy h-RPE cell examined under confocal microscope. Raft domains are BODIPY-FL-Gm1 labeled, showing no clustering. (b) Spatial IFCS spectrum of the image collected from boxed domain of (a). In confocal image (c), the h-RPE cell has been subjected to an apoptotic trigger. Image shows the clustering of raft domain as a consequence. In (d), the spatial IFCS from boxed domain of (c) shows that the clustering leads to a spatial correlation value of ~ 0.35 in cluster density. Wu (unpublished thesis, 2009).

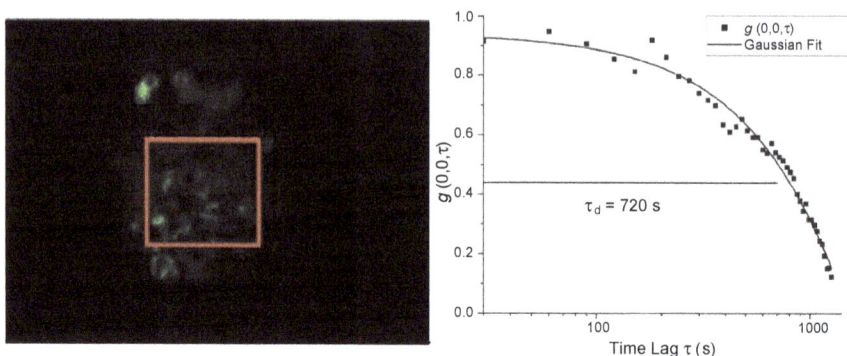

Figure 7.26. Temporal autocorrelation (temporal IFCS) function $g_{ij}(0,0,\tau)$ computed from the selected (red) area in a sequence of images of healthy ARPE-19 cells, taken every 30 seconds. Diffusion characteristic time τ_d, the time when $g_{ij}(0,0,\tau) = g_{ij}(0,0,0)/2$, quantifies diffusion coefficient. $D = w2/(4\tau d) = 0.03\ \mu m^2/s$. Wu and Yeh (unpublished) and Wu (2009).

If instead, we were to sample the time evolution of one of the images, we can calculate the time evolution of each of the points, and in particular, plot a specific single point $(0,0)$ time correlation. Thus

$$g_{ij}(0,0,\tau) = g_{ij}(0,0,0)(1 + \tau/\tau_d)^{-1}. \tag{7.17}$$

In **Figure 7.26**, the temporal IFCS data yields a temporal correlation time, τ_d, for the dynamics of fluorescent signal examined at the $(0,0)$ point of the x–y plot. Clearly different points of the image can be equally followed and statistically averaged.

Thus information from the entire set of images exists and can be displayed by appropriate choice of space or time points. The convenience of this technique is, however, offset by the difficulty of not having enough spatial or temporal resolution due to the camera's intrinsic pixel resolution and recycling rate in time.

7.8   Fluorescence Resonant Energy Transfer (FRET) FCS

In a previous section (Chapter 5), we had discussed the usefulness of the technique called Fluorescence Resonant Energy Transfer (FRET) as a "molecular ruler." If two fluorophores are at a distance short enough for the resonant energy transfer to play a significant role, < 10 nm, FRET intensity from the dipole–dipole energy transfer process is efficient, and accurate measurement of the distance of spacing between the donor and acceptor fluorophores can be achieved. What is also available is the temporal relationship between the donor and the acceptor. If the fluorophore of the acceptor molecule was coming into or moving out from the short $(1/R^6)$ dependence for effective FRET signaling, the time change of the FRET signal of the acceptor molecule is an indication of the time evolution of these two molecules. Often molecular blinking events can be sampled and quantitatively analyzed using the FRET FCS signal. The increase in the FRET signal indicates the position between the donor–acceptor pair is getting short while a decrease in that FRET intensity suggests the reverse — fluorophores are getting further away from each other. In a study by Torres and Levitus (2007), this method has been applied for the measuring of chemical reaction rates. In their study shown in **Figure 7.27**, the two species of a binary chemical reaction are each individually labeled by a fluorescent donor and acceptor, respectively. When the species are nearby spatially, as in a molecular complex, the FRET signal is strong. However, when the reaction is favoring the independent species, the FRET signal is

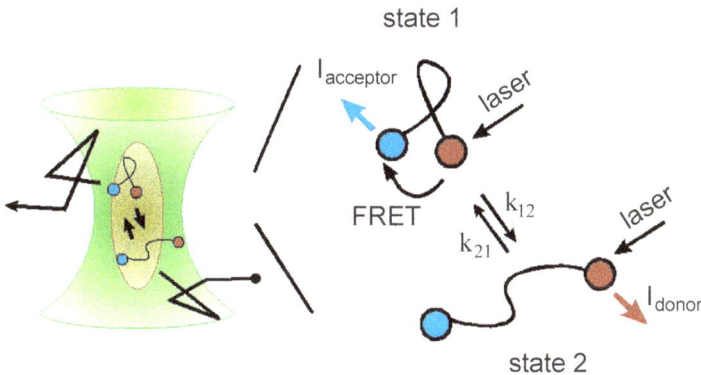

Figure 7.27.   Principle of FRET-FCS. For a two-state system the kinetic rates of interconversion between states 1 and 2 are represented by k_{12} and k_{21}. The biomolecule contains a donor (red) fluorescent probe which is excited directly by a laser. A fraction of the excitation energy is transferred to an acceptor (cyan) molecule by FRET, with a distance-dependent efficiency. A small open volume is created by focusing a laser beam to a diffraction-limited spot and using a confocal pinhole to limit the detection of photons to light originating from regions close to the focal plane (observation volume). Molecules diffuse freely in and out of the observation volume, producing fluctuations in the fluorescence intensity of both acceptor and donor. Fluctuations are due to diffusion and to changes in FRET efficiency due to the conformational dynamics of the macromolecule. Reproduced with permission from Torres and Levitus (2007).

minimized. The FRET-FCS time dynamic measurement becomes a probe to monitor the reaction rates. This technique can be implemented using a combination of two FCS channels, similar to FCCS. The contributions of particle diffusion as a complication to reaction parameters, as well as the usual cross-talk signal as a complication in FRET studies are being worked out.

7.9 Summary

Fluorescently labeled species can be monitored in the time domain to extract the extensive amount of information about the structure and kinetics of the labeled species. The combination of super-resolution microscopy and FCS is definitely not too far on the horizon. We anticipate future studies to couple the spatial localization of ultrahigh resolution microscope with the temporal dynamics that can be probed over many timescales.

References

Axelrod, D., P. Ravdin, D.E. Koppel, J. Schlessinger, W.W. Webb, E.L. Elson and T.R. Podleski. Lateral motion of fluorescently labeled acetylcholine receptors in membranes of developing muscle fibers. *Proc. Natl. Acad. Sci. U.S.A.* 73: 4594–4598, 1976.

Bacia, K. and P. Schwille. Practical guidelines for dual-color fluorescence cross-correlation spectroscopy. *Nat. Protoc.* 2: 2842–2856, 2007.

Berne, B. and R. Pecora. *Dynamic Light Scattering: With Applications to Chemistry, Biology and Physics*, John Wiley & Sons, New York, 1976.

Bulseco, D.A. and D.E. Wolf. Fluorescence correlation spectroscopy: Molecular complexing in solution and in living cells, *Methods in Cell Biology*, Academic Press, New York, 2007.

Carrasco, C., M.S. Dillingham and F. Moreno-Herrero. Single molecule approaches to monitor the recognition and resection of double-stranded DNA breaks during homologous recombination. *DNA Repair* 20: 119–129, 2014.

Chu, B. *Laser Light Scattering*, Academic Press, Inc, 1974.

Chu, B. *Laser Light Scattering: Basic Principles and Practice*, Academic Press, New York, 1991.

Cummins, H.Z., N. Knable and Y. Yeh. Observation of diffusion broadening of Rayleigh scattered light. *Phys. Rev. Lett.* 12: 150–153, 1964.

Fan, S.F., M.M. Dewey, D. Colflesh and B. Chu. Effect of ATP depletion on the isolated thick filament of limulus striated muscle. *Biophys. J.* 52: 859–866, 1987.

Fore, S., T.A. Laurence, Y. Yeh, R. Balhorn, C.W. Hollars, M. Cosman and T. Huser. Distribution analysis of the photon correlation spectroscopy of discrete numbers of dye molecules conjugated to DNA. *IEEE J. Sel. Top. Quantum Electron.* 11: 873–880, 2005.

Fore, S., Y. Yuen, L. Hesselink and T. Huser. Pulsed-interleaved excitation FRET measurements on single duplex DNA molecules inside C-shaped nanoapertures. *Nano Lett.* 7: 1749–1756, 2007.

Fore, S.R. *Development of single molecule optical techniques for the study of recognition and repair of DNA damage.* Thesis (Ph.D.). University of California, Davis, 2006.

Fujime, S., M. Takasaki-Ohsita and S. Ishiwata. Dynamic light-scattering study of muscle F-actin. II. *Biophys. Chem.* 27: 211-224, 1987.

Koppel, D.E., D. Axelrod, J. Schlessinger, E.L. Elson and W.W. Webb. Dynamics of fluorescence marker concentration as a probe of mobility. *Biophys. J.* 16: 1315–1329, 1976.

Korlach, J., P.J. Marks, R.L. Cicero, J.J. Gray, D.L. Murphy, D.B. Roitman, T.T. Pham, G.A. Otto, M. Foquet and S.W. Turner. Selective aluminum passivation for targeted immobilization of single DNA polymerase molecules in zero-mode waveguide nanostructures. *Proc. Natl. Acad. Sci.* 105: 1176–1181, 2008.

Krieger, J.W., A.P. Singh, N. Bag, C.S. Garbe, T.E. Saunders, J. Langowski and T. Wohland. Imaging fluorescence (cross-) correlation spectroscopy in live cells and organisms. *Nat. Protoc.* 10: 1948–1974, 2015.

Kubota, K., Y. Tominaga and S. Fujime. Dynamic light-scattering study of semiflexible polymers: Collagen. *Biopolymers* 26: 1717–1729, 1987.

Levene, M.J., J. Korlach, S.W. Turner, M. Foquet, H.G. Craighead and W.W. Webb. Zero-mode waveguides for single-molecule analysis at high concentrations. *Science* 299: 682–686, 2003.

Liu, B., R.J. Baskin and S.C. Kowalczykowski. DNA unwinding heterogeneity by RecBCD results from static molecules able to equilibrate. *Nature* 500: 482–485, 2013.

Ly, S., F. Bourguet, N.O. Fischer, E.Y. Lau, M.A. Coleman and T.A. Laurence. Quantifying interactions of a membrane protein embedded in a lipid nanodisc using fluorescence correlation spectroscopy. *Biophys. J.* 106: L05–L08, 2014.

Ly, S., J. Petrlova, T. Huser, S. Fore, T. Gao, J. Voss and T.A. Laurence. Stoichiometry of reconstituted high-density lipoproteins in the hydrated state determined by photon antibunching. *Biophys. J.* 101(4): 970–975, 2011.

Magde, D., E. Elson and W.W. Webb. Thermodynamic fluctuations in a reacting system — measurement by fluorescence correlation spectroscopy. *Phys. Rev. Lett.* 29: 705–708, 1972.

Pack, C.G., H. Yukii, A. Toh-E, T. Kudo, H. Tsuchiya, A. Kaiho, E. Sakata, S. Murata, H. Yokosawa, Y. Sako, W. Baumeister, K. Tanaka and Y. Saeki. Quantitative live-cell imaging reveals spatio-temporal dynamics and cytoplasmic assembly of the 26S proteasome. *Nat. Commun.* 5: 3396, 2014.

Riva, C., B. Ross and G.B. Benedek. Laser Doppler measurements of blood flow in capillary tubes and retinal arteries. *Invest. Ophthalmol.* 11: 936–944, 1972.

Samiee, K.T., J.M. Moran-Mirabal, Y.K. Cheung and H.G. Craighead. Zero mode waveguides for single-molecule spectroscopy on lipid membranes. *Biophys. J.* 90: 3288–3299, 2006.

Sauer, M. Localization microscopy coming of age: From concepts to biological impact. *J. Cell. Sci.* 126: 3505–3513, 2013.

Saxton, M.J. Modeling 2D and 3D diffusion. In: A.M. Dopico (ed.) *Methods in Membrane Lipids*, Totowa, NJ: Humana Press, 2007.

Saxton, M.J. Single-particle tracking. In: *Fundamental Concepts in Biophysics*, Humana Press, New York, 2007.

Schwille, P. and E. Haustein. Fluorescence correlation spectroscopy: An introduction to its concepts and applications. In: D. Mcgavin (ed.) *Biophysics Textbook Online*, Biophysical Society, Rockville, MD, USA, 2004.

Torres, T. and M. Levitus. Measuring conformational dynamics: A new FCS-FRET approach. *J. Phys. Chem. B* 111: 7392–7400, 2007.

Warshaw, D.M., G.G. Kennedy, S.S. Work, E.B. Krementsova, S. Beck and K.M. Trybus. Differential labeling of myosin V heads with quantum dots allows direct visualization of hand-over-hand processivity. *Biophys. J.* 88: L30–L32, 2005.

Wiseman, P.W. and N.O. Petersen. Image correlation spectroscopy. II. Optimization for ultrasensitive detection of preexisting platelet-derived growth factor-beta receptor oligomers on intact cells. *Biophys. J.* 76: 963–977, 1999.

Wu, H. *Study of membrane dynamics with biophotonic techniques.* Thesis (Ph.D.), University of California, Davis, 2009.

Xu, W., W. Liu, J.F. Schmidt, W. Zhao, X. Lu, T. Raab, C. Diederichs, W. Gao, D.V. Seletskiy and Q. Xiong. Correlated fluorescence blinking in two-dimensional semiconductor heterostructures. *Nature* 541: 62–67, 2017.

Yeh, Y. and H.Z. Cummins. Localized fluid flow measurements with an He-Ne laser spectrometer. *Appl. Phys. Lett.* 4: 176–178, 1964.

Yildiz, A., J.N. Forkey, S.A. Mckinney, T. Ha, Y.E. Goldman and P.R. Selvin. Myosin V walks hand-over-hand: Single fluorophore imaging with 1.5-nm localization. *Science* 300: 2061–2065, 2003.

Chapter 8

Photonics in Medicine

The science (or art) of medicine is devoted to methods and approaches toward the bettering of the health of the human body (and mind) by whatever means possible. From time immemorial, approaches have included the worshipping of spirits and the extraction of the presumed diseased parts. Medicine has come a long way, but we are just beginning to understand the complexity of the human body and mind. Through our own current understanding, we are also starting to appreciate some of the reasons for those early methods of medical practice.

There are two aspects to medicine: preventive and curative. In prevention, the goal is to ensure that a well body stays well. That means we need to fully understand what that well body is and how it functions to maintain wellness on its own. This is a tall order since we are now learning that the human body is a very complex piece of machinery, with so many intricately interconnected components. Equally a monumental order is the cure for a disease. We are starting to see how some of the disease recognition avenues operate in the healthy body in today's biology studies. Many of the complex interplays we have discussed in our chapter on cellular biology foretell the even more complex nature of how a body deals with invading diseases of one type or another. Gene repair such as telomere management, cellular regulation including apoptosis and autophagy, and membrane signaling for downstream action involving G-protein coupled receptors (GPCR) approaches are all methods that a cell effect self-regulation. Fortunately, there are some commonalities across all types of diseases. These common factors provide us with new approaches to deal with diseases. These include:

(1) How does a healthy body become diseased?

(2) How to see and localize, with as much clarity and definition as possible, the existing diseased state.

(3) How this particular disease is causing harm to a healthy body.

Optics is starting to play significant roles in all of these avenues.

8.1 Visualization and Validation

8.1.1 *Imaging with microscopy and endoscopy*

One of the earliest applications of medical imaging is the use of X-rays to examine the possibility of bone fractures. Since the human body has evolved to a state where the structural elements, bones, and cartilages are well shielded by tissues and skins, even the gross fracture of a large bone is not necessarily visible to the naked eye from the outside. With the discovery of X-rays that are energetic enough to penetrate the protective tissues but are absorbed selectively by the bony structures, early X-ray images already show the existence of these structures, with any of its potential issues related to deformities or breakages. Once seen, a medical practitioner of that particular expertise (orthopedics) can provide the best approach to remove, shape, or simply reset the bone. The extension of this technique to imaging softer tissues including the cartilages and abnormal joint growths has helped physicians to deal with arthritic joints and athletes' torn ligaments. Even softer (lower energy) X-rays are now used to image large organs of the body so that diseases of the lungs, liver, pancreas, and other smaller organs can be visualized. The entire field of X-ray *computed tomography* (CT) is based on the use of X-ray dyes that are dispersed into the body and then imaged using modern, softer X-rays, producing sectional images of the body (tomographically) with sufficiently high contrast to be recognized.

The main concerns with using X-rays, which has the potential for ionizing and otherwise damaging the body's cell, is excess dosimetry that can cause unexpected radiative damage to the healthy tissues. For example, a typical chest X-ray CT image delivers about 7 milliSievert (mSv). For comparison, the ambient, naturally occurring background X-rays deliver about 3 mSv/year to a human body, which we assume to be safe. So on average, one chest X-ray examination every two years is roughly an allowable dosage. The damage created by X-rays can be in the form of inducing DNA damage within the cells nearby thus causing a cellular mutation that could lead to abnormal cell growth or eventually cancer. Due to this concern, as much as possible, the CT field searches for newer, less invasive sources of radiation and less interactive contrasting agents. In the meantime, biomedical engineering focuses on other means of imaging that are less invasive than X-rays. These include positron emission tomography (PET) and acoustic (ultrasonic) imaging.

With respect to the use of optical wavelength light for probing the body, one immediately encounters the most formidable issue of tissue transparency. An ingenious solution to partially resolve this complication is the development of Optical Coherence Tomography.

8.1.2 *Optical Coherence Tomography (OCT)*

A reason for the difficulty of using the longer wavelength of visible light to conduct imaging studies of a body is light scattering. So many of the components of tissues are of such sizes that they scatter visible light. Thus any depth sampling will encounter the phenomenon of multiple scattering. It is well-known that multiple scattering destroys both the phase and polarization of light, rendering the desired information from the single scattering processes lost. *Optical Coherence Tomography* (OCT) is one method that can overcome some of this difficulty.

As an example, human skin has a mean free path for visible light of about 5 μm. This is significant because if one wants to discern a feature that is just a single millimeter (1 mm) below the skin surface, the light will have undergone many, many scattering processes. This has the effect of (a) the target receiving only a very small amount of the incident, focused light, and (b) the light scattered from the target has even less chance of directly getting back to the detector on the surface of the skin. The result is a loss of intensity of the needed signal, hence signal-to-noise ratio (S/N), and even more importantly, a lack of spatial and polarization resolution. The idea of OCT is to pick up that small number of singly scattered backscattered light from the target in the presence of all the multiple scattering events (noise).

OCT utilizes "coherence gating" to discriminate the weak, single scattered light from the background of many multiply scattered light. This method is based on the use of a Michelson interferometer, but here, the light source is one that is broadband and only partially coherent.

Let us first examine one of these interferometers. In **Figure 8.1**, we show the basics of the Michelson interferometer. This has a light source, a beam splitter, a reference arm with a movable mirror, and a signal arm,

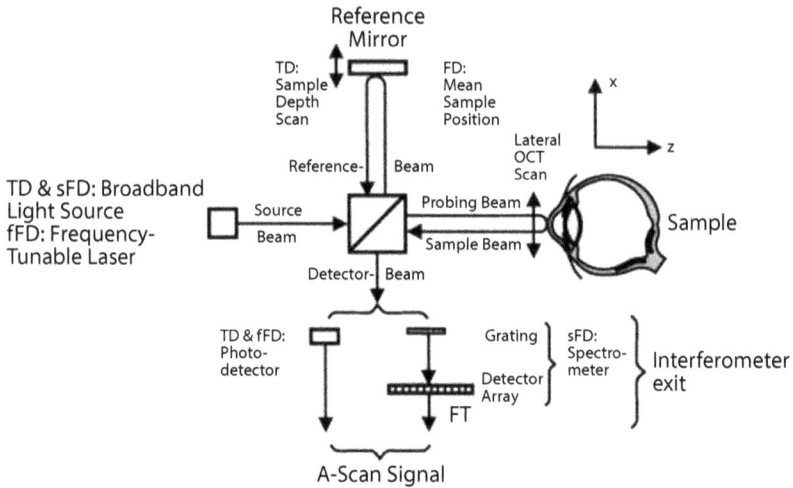

Figure 8.1.   Standard time-domain (TD) and Fourier-domain (FD) Michelson interferometer configurations used in OCT. Note different functionings of the reference mirror and different signal detection at the interferometer exit. FD-OCT techniques use either a broadband light source (sFD) or a frequency-tunable laser (fFD). Reproduced with permission from Fercher (2010).

here represented by a sample, which is an eye. The recombined beam will produce an interferogram on the detector beam for interferometer exit signal. Note also in Figure 8.1 the reference to time-domain (TD) sampling versus frequency-domain (FD) sampling modes. Both approaches have been used in OCT imaging studies. As suggested in this figure, TD sampling is mostly used for depth studies whereas FD sampling allows ease of lateral positioning measurement. Sensitivity issues in detection currently favor the FD approach.

Consider first if we had used a coherent laser source. In that case, the resulting interferometric intensity as measured by a square law detector, due to the ability of the coherent light source to interfere constructively and destructively in a periodic manner continuously, will lead to a constant envelope value, no matter what depth the signal is coming from. If instead, the excitation light is only partially coherent, then the interferometer pattern takes on the shape of **Figure 8.2**, where interference is seen only *within* a coherence length, L_c, where $L_c = (c/n)\tau_c$, and the coherence time of the source is τ_c. This idea means that if the light source is of short coherence length, any movement of the reference mirror beyond the coher-

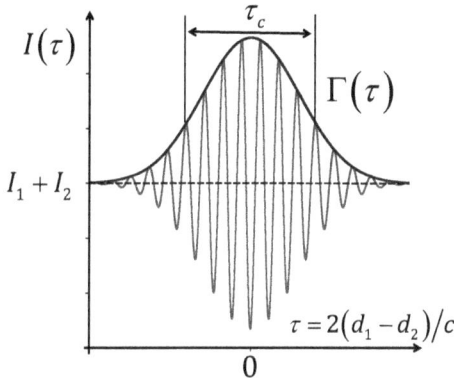

Figure 8.2. Temporal interferogram from a light source of coherence time τ_c. I_1 and I_2 are the two intensities from the two arms. d_1 and d_2 are the distances that light traverses in the respective arms.

ence length will negate the interference pattern. In Figure 8.2, the detected time domain interferometric signal is shown as a series of interferometric fringes that fade away rapidly when the time of arrival of the two signals is longer than the source coherence time, τ_c. Here, d_1 and d_2 are the optical distances from the two paths of the Michelson interferometer. Thus, should the sample arm produce backscattered light that has any changes in phase, such as depth change, a change in the interferometric intensity will ensue. High axial spatial resolution is then possible because the fringes are observed only when the optical path length difference between the sample arm and the reference arm is less than the coherence length L_c. So depth resolution limited by the source coherence length can be discerned.

Mathematically, the local, single point signal is given by

$$I = I_s + I_r + 2\sqrt{I_s I_r}\Gamma(\Delta\tau)\cos(\omega\Delta\tau) \tag{8.1}$$

where

$$\Gamma(\tau) = E^*(t) * E(t + \tau). \tag{8.2}$$

$\Gamma(\tau)$ is the time correlation function of the total signal, with $E = E_r + E_s$. Here, subscript s is the sample arm signal, and r is the reference arm signal. Hence the local signal intensity measured by I is a representation of the degree of phase coherence between the reference and signal arm. By sweeping the focal point of the source in that plane, the different signals obtained via a complete lateral sweep of the points being probed will report the difference in backscattered light's arrival time to the detector. Note that difference in either physical length or optical density will yield a change in the interferometer intensity signal, up to the coherence length of the low coherence source (LCS).

As in any heterodyning device, the intensity of the signal measured in time t at any lateral position (x, y) is given by

$$I_d(\tau) = I_r + I_s + \Gamma_{\text{OCT}}(\tau). \tag{8.3}$$

Since the intensities of the individual arms of the interferometer, reference I_r, and signal I_s, are each essentially non-varying, the observed change comes only from the interference generating OCT pattern

$$\Gamma_{\text{OCT}}(\tau) \propto \exp\left[-4\ln 2\frac{\tau^2}{\tau_c^2}\right]\cos(\omega_0\tau). \tag{8.4}$$

Here, the incident light temporal distribution is assumed Gaussian, with τ_c as the coherence time of the light source. Often it is more useful to convert this relationship to one exhibiting path length differentials, ΔL.

$$\Gamma_{\text{OCT}}(\tau) \propto \exp\left[-4\ln 2\frac{\Delta L^2}{L_c^2}\right]\cos\left(4\pi\frac{\Delta L}{\lambda_0}\right) \tag{8.5}$$

where λ_0 is the incident wavelength in vacuum, L_c is the coherence of the light source, and ΔL is the axial displacement distance. What one sees is that the envelope of the oscillation function is a strong function of the coherence length. So by measuring the fringe pattern and varying over the lateral position of the sample, one can map out the optical density of spatial regions up to the resolution of this instrument. Using ultra-broad bandwidth light sources, it has been possible to achieve axial resolution of ~ 1 μm.

One area of major OCT application is in the field of ophthalmology. Because the eye is the intrinsic optical device of the human body, it provides a window for light to enter and be detected and later interpreted by the brain. The detection takes place on the retinal surface that is full of light receptor cells such as the rods and cones. One major disease associated with the retinal surface is macular degeneration. In various forms, the basic idea is that the retina has detached itself from the connecting cells, hence rendering that region of the retina non-responsive to light, as far as the brain is concerned. One highly definable retinal defect is called the *macular hole*, shown in **Figure 8.3** when viewed with modern clinical OCT apparatus. It is clearly seen that the healthy macula region (Figure 8.3a) has a continuous retinal surface as opposed to that seen in Figure 8.3b. The OCT image

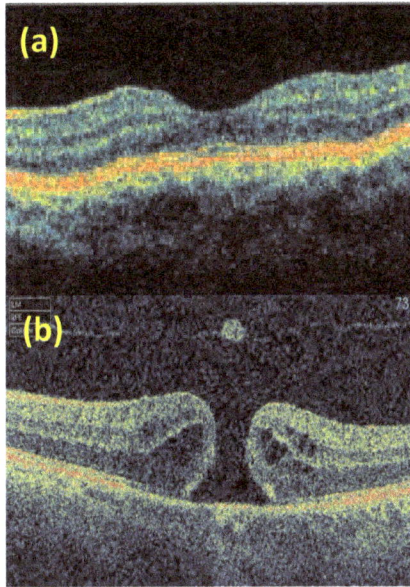

Figure 8.3. (a) Single OCT scan line from a healthy subject and (b) in patient with a macular hole. Panel (a) is reproduced with permission from Van Velthoven *et al.* (2007).

shown here is marking a retinal hole, seen as a local detachment of retina with a width ~ 200 μm as an opening. The image also shows the depth of the retinal tissue associated with the hole. This optical tool allows an ophthalmologist with retina specialty to design treatments to repair this region and possibly restore an individual's eyesight.

Often in the early stages of diseases, the changes in the tissue pattern from that of normal tissues are very minimal. Thus it is not easy to extract meaningful data from the conventional, intensity-based OCT sampling protocol. Many very advanced approaches have allowed for more specific details of the diseased state to be measured. These include the use of functional OCT (F-OCT), of which Doppler OCT (D-OCT), which tracks the movement of the species based on the Doppler effect, is one (Chen *et al.*, 1999). Polarization-sensitive OCT (PS-OCT) is another approach to increase selectivity because often the diseased state exhibits some molecular randomness that did not exist in the normal state.

In a recent study by Qu *et al.* (2017), F-OCT is coupled to an acoustical transducer to probe the change in cardiovascular vessel elasticity upon the buildup of plaque. The authors found that OCT has higher spatial reso-

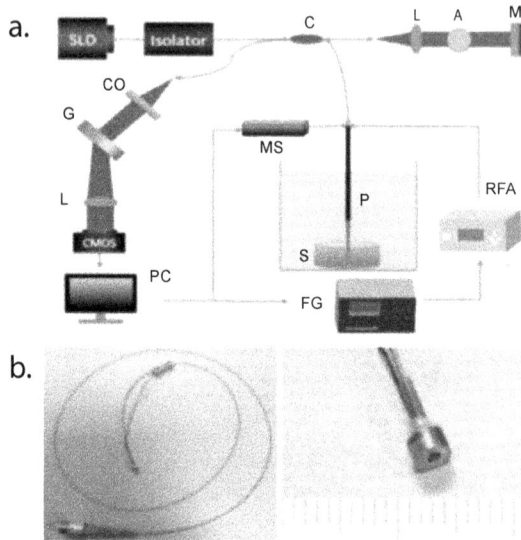

Figure 8.4. Optical fiber-based ARFI-OCT system that is used to monitor intra-vascular plaque tension in cardiovascular vessels. Shown here are a. The OCT device using grating to produce Fourier-domain detection. RFA is the radio frequency amplifier for the RF driver (FG) of the acoustical transducer. b. Device shown as the fiber optical element including a ring acoustical transducer (3.5 mm diameter) and optical elements within. Reproduced under Creative Commons CC-BY license from Qu *et al.* (2017).

lution than acoustical radiation force imaging (ARFI) alone. They used a fiber-optical scheme shown in **Figure 8.4**. In this scheme, a superluminescent diode (SLD) with a 150 nm bandwidth was the light source. Such a fiber-optical system allows the reference and sample signals to be combined using a coupler and the correlation signal sent onto a grating for FD data processing. It is interesting to note the size of these components as shown. Such a probe can be incorporated into an *endoscope* and inserted into the part of the vessel/chamber for *in vivo* monitoring of the condition of the tissue. In this case, it is shown that the instrument is sensitive to changes in elastic tension of the surface, such as that which can be found in the buildup of plaque in a carotid artery. This study shows that these changes of elasticity of the tissue can be monitored by a change in the Young's modulus of elasticity as measured by F-OCT.

Second harmonic generation (SHG) process is another one that will detect well-defined differentials between normal and diseased states. It is well-known that molecular organization exhibiting chiral features (helices) in normal tissue molecules will have decreased organization in the diseased sample. Thus second harmonic Optical Coherence Tomography (SH-OCT) has been used to detect early changes in the structural elements from skin and tendons. In a recent review by Chen and Tang (2015) on the topic of OCT, the authors reviewed the instrumentation that allowed the simultaneous acquisition of primary OCT and SH-OCT images from the same sample. A schematic diagram of one of their earlier apparatus is shown in **Figure 8.5** (Su *et al.*, 2007).

Here, several features need attention. First of all, in order to be effective as a broadband light source, the incident femtosecond fiber laser (FSL) has to have its beam prepared. In the scheme shown here, that feature is achieved by a single-mode fiber stretcher. The SH reference beam is created

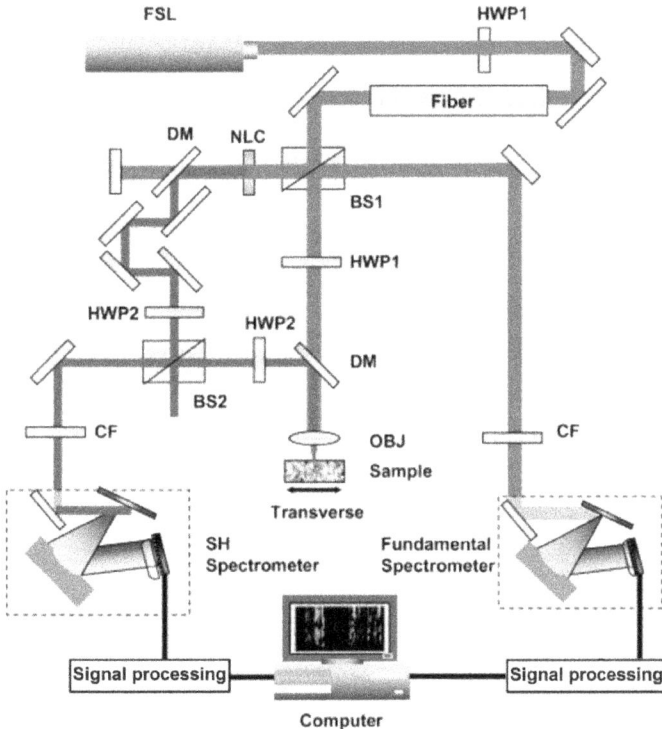

Figure 8.5.   Schematic of the FD SH-OCT experimental system. FSL: femtosecond fiber laser; HWP: half-wave plate; DM: dichroic mirror; NLC: nonlinear crystal; OBJ: objective; BS: beam splitter; CF, color filter. Reproduced with permission from Su *et al.* (2007).

by the use of a nonlinear crystal. We note that at the beam splitter, the reference beam is directed to generate its own reference SHG reference signal. At the subsequent dichroic mirror, the primary reference is deflected to interfere with the sample's primary backscattered signal whereas the SHG signal from the sample mixes only with the SHG reference beam. In this frequency-domain (FD) scheme, the authors achieved better sensitivity than the time-domain OCT signals. For the weaker SHG signal, this feature is important. The FD conversion is achieved by independent gratings for the primary and second harmonic signals.

Both sets of OCT images are compiled on the computer. Data obtained from the fish scale of salmon in this study (Su *et al.*, 2007) are displayed side by side as images from intensity-only OCT and from SHG-OCT (**Figure 8.6**). It is clearly seen in comparison that whereas the intensity OCT emphasizes the optical density of the regions, SHG-OCT is dominated by the collagenous hence helical or chiral domains. In combination, such images are invaluable for identifying more detailed structures, or damages to molecular structures in diseased states.

It is of interest to note that SHG-OCT also has the same depth sectioning capability as other nonlinear excitation/imaging modalities.

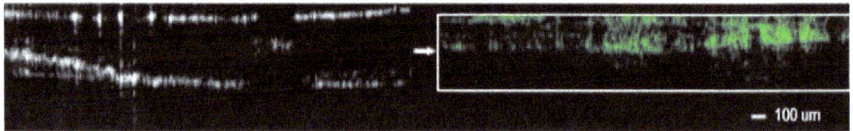

Figure 8.6. Fish scale OCT and SH-OCT images. On the left-hand side is the fish scale OCT image; on the right-hand side, the SH-OCT image is shown. Reproduced with permission from Su *et al.* (2007).

8.1.3   *Optical endoscope*

The endoscope is, as its name suggests, a device that allows for monitoring of the internal domains of organs and vessels. In the main, these are also optical fiber devices set up to carry out different functions of the normal tabletop optical systems. The backscattering collection called for in the OCT configuration in the previous section is but one of the many arrangements. We have already seen the use of optical fibers as an OCT device conduit. Tailoring the small tips for entry into closed quarters renders these endoscopes. The primary requirements of the endoscope are three.

(1) Very thin and flexible so as not to be intrusive toward the function it is designed to monitor. This may be a single fiber with adequate cladding to shield from the environment, and a tiny *gradient index* GRIN lens to allow adequate focus or collection of light. Single mode fibers are special fibers with cross-sectional diameters less than the optical mode dimension, rendering them capable of carrying only those waves that meet the proper spatial boundary conditions of sustaining $\frac{1}{2}$ wavelength. Such fibers are used when single transverse mode demand is beneficial for S/N.

(2) Not only does this optical fiber have to carry the incident light to the location for the specific monitor, the return signal, backscattering or fluorescence, must be captured and processed. In the case of the OCT configuration, the return backscattered light is sent into the mirror to be optically mixed with the reference beam. In the fluorescence situation, that light must be sent into the spectrometer or a wavelength passband filter and subsequently to a sensitive detector. Simultaneously the output must be equipped with an eyepiece or camera so that the image can be visualized in real time.

(3) The exterior surface of the endoscope must be totally inert with respect to the path it is designed to take. Furthermore, this surface and the entire endoscope must be able to undergo rigorous sterilization protocols and be totally disposable upon use.

In recent innovations, optical fiber endoscopic probes equipped with OCT capabilities have been used to capture multi-modality images from tissues. In this concept, in order to have multi-modality available at the tip of the fiber, the fiber is constructed with double cladding. The inner, small diameter fiber is the single mode fiber used to generate the standard OCT image. The surrounding clad is a second layer of optical fiber that is multi-modal and has high NA to capture fluorescence or absorption signals in precise locations. In this configuration, there is the need to couple the returned incoherent signal from fluorescence via a broadband coupler section. Boudoux and her group (De Montigny *et al.*, 2015) have achieved this to measure the fluorescence signal distribution derived from labeled antibodies in deep tissue locations as designated by the OCT image. This looks like a promising technique once the background autofluorescence can be eliminated adequately.

In a recent review of this field, Wilson and Borel (2016) discussed the use of nanoparticle-enabled optical endoscopes for higher specificity and better accuracy. We shall discuss nanoparticles in more detail in the next section on biomarkers. Suffice to say in this section that advances in nanoparticle (NP) technology has made these particles an essential component of medical diagnostics and highly focused therapeutic delivery processes.

8.1.4 *Adaptive optics microscopy*

The idea of adaptive optics is to create a lens that can compensate for the distortions the optical encounters during its optical path either onto the sample of interest or from the sample to the detector. The use of adaptive optical corrections was primarily focused in the field of astronomy, where distant objects emitting light will traverse through a much stellar medium with optical distortions. A National Science Foundation (NSF) Center for Adaptive Optics was created to examine this topic. Quickly, an extension of the methods to the field of biological imaging became recognized as being equally important. A review of acousto-optical OCT (AO-OCT) by Jonnal *et al.* (2016) highlights the wide range of bioimaging accomplished by the use of adaptive optical lenses to compensate for the dense material variations in index of refraction in the path of the OCT microscope. Essentially, the use of liquid crystal lens or deformable mirrors that are feedback coupled to the optical path (wavefront) monitoring provides the means to minimize optical wavefront distortions, rendering a better-focused beam, hence higher achievable transverse spatial resolution. As adaptive optics technology advances in the future, the range of capability will improve. Adaptive optical corrections are not limited to label-free OCT imaging methods. Ji (2017) discussed the use of adaptive optical methods to higher resolution fluorescence microscopy. The goal is to improve ultrahigh optical resolution microscopy with even higher lateral spatial resolution through the incorporation of this feature.

8.2 Biomarkers

Biomarkers are those signatures that will define a particular species or a particular pathway intermediate. The desirable feature of a biomarker is that each marker shows a specific process related to the species of interest as postulated or predicted. Biomarkers are keys for identifying species and processes related to diseases. This, however, also means that in order to find that important biomarker, it is necessary to carry out a significantly large number of experiments to positively show the unique existence and specificity of this marker molecule. In many ways, photonics tools are ideal for the search, identification, and possibly even destruction of certain important biomarkers for diseased states of a cell. Indeed biomarkers are necessary for showing the different pathways for the biochemical processes in normal cellular functions as well.

An example showing the need for meaningful biomarkers is in the case of identifying the dynamics of intracellular Ca^{++} ions. Ca^{++} is necessary for cellular motility. It is well-known that Ca^{++} is a key catalyst in the binding reaction between myosin and actin. Ca^{++} is required by the molecule troponin so it can be properly oriented to receive the myosin and initiate molecular force generation. However, a faulty Ca^{++} concentration at the needed time sequence of the overall excitation-contraction cycle will render the downstream contractile system non-functional. So one way to track the timing of Ca^{++} release is to monitor the time coordination of the release of these ions from the sarcoplasmic reticulum (SR), a sequester reservoir for Ca^{++} ions. How to visualize calcium currents is through fluorescent dye molecules Fura, or another one called Indo. These are dye molecules that require the chelating of Ca^{++} for it to stabilize its own structure, thus allowing its excited state to emit a fluorescent signal. Without chelation, the molecule is not a strong fluorophore.

The Fura-2 dye molecule is structurally shown in **Figure 8.7**. The double carboxyl groups allow for the binding of two Ca^{++} ions per Fura-2

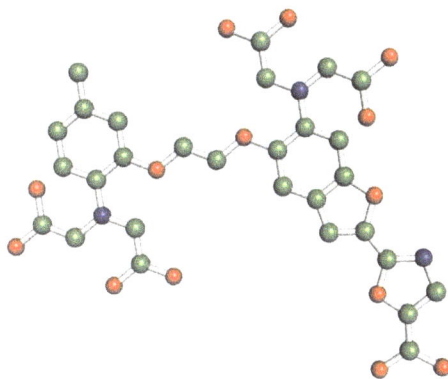

Figure 8.7.   Chemical structure of Fura-2 dye molecule. Figure produced using PyMol (DeLano, 2002) and ChemSpider ID 51442.

molecule. The stabilizing of the structure absorbs around 380 nm and emits at ~ 500 nm.

Such a probe, monitoring the product of SR release of free Ca^{++} by the Ca-ATPase, is a signature of the activity of the protein, Ca-ATPase. A well-functioning Ca-ATPase provides the needed Ca^{++} ions so that the downstream contractile apparatus can carry forth. So the fluorescent dye has acted as an indirect biomarker for protein activities both upstream and downstream.

In our study of the activity level of the RecBCD helicase, we had used as a biomarker the fluorescent dye molecule that is an intercalating dye for dsDNA. In that case, as we had shown and described in Chapter 4, the change in fluorescent intensity between the intercalated dye and that of the free dye is again enormous, thus allowing the dye, YOYO-1 to be used as a biomarker for the functioning test of the protein RecBCD.

An extension of such functional biosensors is a recent study by Mo *et al.* (2017). In their study, plasma membranes of living cells were labeled so as to exhibit the activity of protein kinase A (PKA) in microdomains on the membrane via a fluorescent biosensor that detects this specific biochemical activity. Applying *stochastic optical fluctuation imaging* (SOFI) method, spatially resolved PKA reaction sites were identified in these living cells.

In the field of medicine, biomarkers are all important for the physician to capture the onset of a specific disease at its earliest indication. Among the diseases that are seeking critical biomarker molecules are cancers, cardiovascular diseases, diabetes, and other endocrine diseases, and neurological dysfunctions due to Alzheimer's or Parkinson's diseases. The goal in the search for a biomarker is to be as specific as possible in the ability to identify a particular marker with as little of the indicator marker as possible. Many laboratories around the world focus on specific disease and their pathways. It is from that focus that one is able to come up with a set of irrefutable "must-be-there" molecular indicators that can be used as the biomarker for that specific disease. Once the molecule is determined to be a validated biomarker, the next step in biomedical research is to develop a technique that will allow such a molecule to be detected at its earliest (lowest) level possible. Finally, it is the job of the biomedical engineer to come up with an instrument that will detect such a marker molecule in the shortest time, *in situ* and real time if possible, for as low a cost to the user/patient as possible. These are very general prescriptions for the search for biomarkers. We will highlight just a few approaches.

8.2.1 Use of the microRNA detectors

MicroRNA (μRNA) is a recently discovered class of non-coding RNAs that play key roles in the regulation of gene expression. They seem to function at the post-transcriptional level; it is suspected that these molecules may fine-tune the expression of as much as 30% of all mammalian protein-encoding genes. μRNAs are small 22 nucleotide chains that typically encode at various terminal points of a gene. The mature μRNA is incorporated into a ribonucleic acid particle to form the *RNA-induced silencing complex*, RISC, which mediates gene silencing.

μRNAs have been shown to be involved in a wide range of biological processes such as cell cycle control, apoptosis and several developmental and physiological processes including stem cell differentiation, hematopoiesis, hypoxia, cardiac and skeletal muscle development, neurogenesis, insulin secretion, cholesterol metabolism, aging, immune responses, and viral replication. In addition, studies also suggest that μRNAs play a vital role in the differentiation and maintenance of tissue identity. μRNAs have also been implicated in a number of diseases including a broad range of cancers, heart disease, and neurological diseases. Thus μRNAs are intensely studied as candidates for diagnostic and prognostic biomarkers and predictors of drug response.

Due to their relatively simple structure, research groups have designed specific methods to identify specific μRNA through the use of complementary base pairing as identifiable biomarkers. In surface plasmon resonance studies, the use of base-pairing for specific μRNA is being developed to create hybridization and consequently to be used as a label-free biomarker where the sensitivity of the device may allow for nano-molar of the μRNA to be directly detected.

8.2.2 From antibodies to SHALs and aptamers

Once the biochemistry and medical laboratory has determined a specific biomarker for a specific disease, in order to achieve its detection at a very low concentration, typically an indicator molecule is attached to this biomarker. One of the robust technologies in biomedicine is the manufacturing of specific antibodies. These are rather large protein molecules that have molecular groups that can be designed to tag the biomarker molecule specifically.

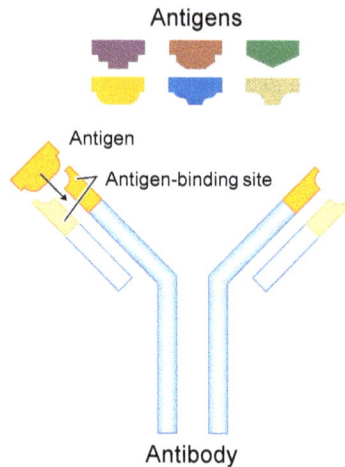

Figure 8.8. Cartoon representation of antigen-antibody complex. Reproduced under wiki-commons license.

The earliest discovery is the antibody (Ab) molecule. Within the immunoglobulin superfamily, the antibodies are made up of glycoproteins and follow a general architecture with two large heavy-chains and two small light-chains (**Figure 8.8**).

The specificity of the antibody comes from the terminal groups of the heavy- and light-chain terminal groups. As shown in the diagram, this terminal group can be tailored to accept only specific antigens (biomarker or other proteins). To be used as a biomarker, antibodies can be labeled to serve as a beacon for the presence of the antigen molecule. Note that since the antibody is a relatively large molecule in its own right, sometimes the binding of a small target biomarker to this large Ab can hinder the detectability due to its physical size. So research has focused on the decreasing size of the ligand. One approach is the Small High-Affinity Ligand (SHAL) molecule. Highly specific but small, stable ligands have been manufactured to focus on these biomarker molecules for labeling but not so much of presenting a hindrance in function or motility. SHALs can be produced *en masse* and of various (**Figure 8.9**) detailed structures. This library of SHALs can then be tapped to find the binding match against a particular biomarker molecule. Once this has been identified, one can develop labels to identify this as a small but specific biomarker. Balhorn *et al.* (2007) have developed SHALs to be used as cancer markers, specifically focusing on lymphoma.

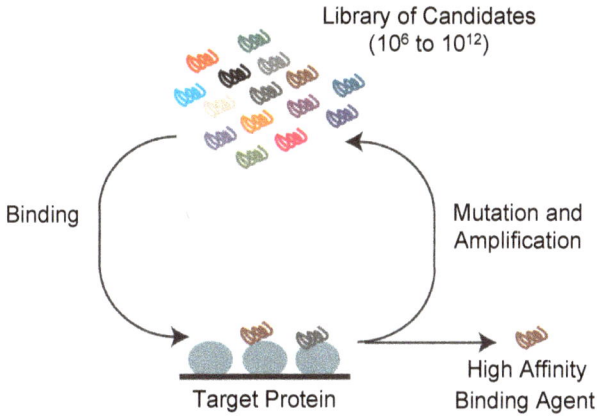

Figure 8.9. Process of traditional *in vitro* ligand discovery. Reproduced from Diehnelt *et al.* (2010) under the terms of the Creative Commons Attribution License.

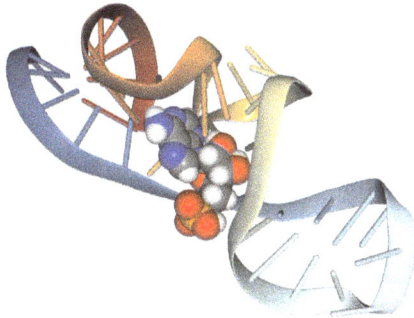

Figure 8.10. Structure of ATP-binding RNA aptamer in complex with AMP. Figure generated using PyMol (DeLano, 2002) with pdb (1raw).

Aptamers are made of single-stranded (ss) RNA or DNA molecules that have the ability to bind with high selectivity to a range of proteins or peptides (**Figure 8.10**). As we had seen in our discussion about ssDNA or ssRNA, the hallmark feature of these is their immense flexibility. Furthermore, base pairing between the complementary bases can occur to create hairpin double strand domains, creating some rigidity in an otherwise very flexible molecule. Thus these molecules can assume a variety of shapes due to their propensity to form helices and single-stranded loops, giving them the versatility in binding to diverse target molecules. Aptamers are used as sensors, and possibly even as therapeutic tools to regulate cellular processes or to guide drugs to their specific cellular targets. Aptamers are manufactured; their specificity and characteristics are tailor-made by the required target's tertiary structure.

It should be very clear from our little discourse on the topic of biomarkers that the optical scientist/engineer will have to be working with focused biochemists and biophysicists to realize the specific goals in the many different biomarker research efforts. The field's progress requires truly interdisciplinary efforts.

8.2.3 *Multiplexed microbead assays — xMAP technology*

The xMAP technology allows the measurement of multiple of analytes (multiplex) simultaneously in a high-throughput fashion within a single biological sample. The multiplex assays can detect nucleic acid and immunoassay formats, enabling simultaneous detection and typing of bacteria, fungi, parasites and viruses, and also antigen or antibody interception (Krishnan *et al.*, 2009, 2011; Reslova *et al.*, 2017).

The principle of multiplex microbead technology is based on the concept of a liquid (suspension) array. Microspheres are dyed with different fluorophores. These surfaces of the fluorescent microbeads constitute the platform for specific binding reactions. In addition to these reactions, the presence of a bead-bound analyte can be detected with a conjugate coupled to a reporter fluorochrome. These sets of microbeads are embedded with precise ratios of red and infrared fluorescent dyes for the classification of their respective bead sets. Each set of microbead is identified with a fluorophore so that multiple sets of microbead can be mixed to form an array of beads and differentiated with a unique spectral signature. By including a third dye into the design, the technology can be expanded up to 500-member microsphere sets (Dunbar, 2006; Dunbar *et al.*, 2003). The bead-bound analyte is detected with a conjugated antibody (e.g., biotin conjugate) allowing subsequent binding of a reporter fluorochrome (e.g., streptavidin-conjugated phycoerythrin — SAPE). This immunoassay produces a variable amount of the green fluorescent reporter dye, which is proportional to the number of analytes bound to the surface of each microbead. Data can be connected to a data acquisition system that enables the instrument to quantify the green, orange (or infrared), and red fluorescence of each microbead using independent detectors. A red laser excites the dye molecules inside the bead and classifies the bead to its unique bead set, whereas the green laser quantifies the assay at the bead surface thus allow-

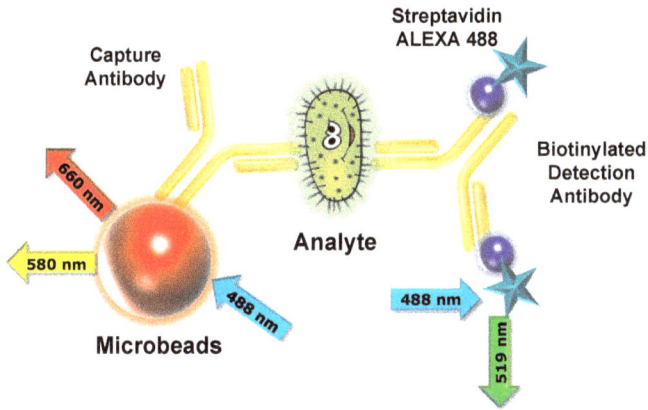

Figure 8.11. Basic principles of microbead sandwich immunoassay. Microbead is covalently coupled with capture antibody. Proteins captured on an antibody microbead are detected by tagged detection antibodies. Inward and outward arrows represent the excitation and emission beams, respectively. Each bead class can be coated with a different capture antibody specific for a single antigen. Upon incubation with samples, the beads form an antibody-antigen complex. This is followed by the addition of biotinylated detector antibodies. Incubation with streptavidin linked fluorescent reporter, R-Phycoerythrin (SAPE), completes the "antigen sandwich." Beads are interrogated one at a time. A red laser classifies the bead, identifying the bead type. A green laser quantifies the assay on the bead surfaces so that only those beads with a complete sandwich will fluoresce in the green, and the signal is a function of antigen concentration.

ing the beads to interrogate. The reporter fluorescence is green (such as FITC (Fluoroscein Isothiocyanate) or Alexa 488), while the Luminex system uses the SAPE complex, an orange (or infrared) fluorescence reporter that is measured by the YAG laser (excitation at 532 nm) (**Figure 8.11**). In the laboratory of Kit Lam, the use of combinatorial chemistry in combination with a one-bead-one-compound technology allowed for rapid screening of millions of chemicals at one time to identify those that bind to diseased cells. This tool is advancing the early detection and precise delivery of treatments for brain, breast, prostate, pancreatic, lymphoma and other cancers (Liu *et al.*, 2017).

8.2.4 *DNA microarrays*

The DNA microarray technology broadly encompasses several techniques that allow multiplexed quantitative determination of gene expression. *In vitro* amplification of nucleic acids by real-time polymerase chain reaction (RT-PCR) enabled the development of a DNA microarray approach that measures the relative concentration of nucleic acids sequences by the detection of the hybridization events. With the advent of results from genome sequencing, the DNA microarray technology utilizes the sequence information to provide information from transcriptional response during a particular physiological stimulus to comprehensive developmental processes. In a typical DNA microarray technology multiple *probes*, individual nucleic acid sequence are immobilized on a solid support (glass slide or bead), and a mixture of fluorescently labeled *targets* (nucleic acid sequence from a sample) is applied as a solution onto the array for hybridization. The subsequent hybridization event is quantified at the specific probe location known as *spot* or *feature* (Brown and Botstein, 1999). The DNA microarray format can be broadly classified into three types: (a) *In situ*-synthesized microarrays; (b) high-density bead arrays; and (c) electronic and suspension bead microarrays. In all these formats, a successful hybridization between the probe and target will have an increase in fluorescence intensity at the spot (feature), which can be quantified.

An overview of the steps in a typical DNA microarray experiment is illustrated in **Figure 8.12**. The total pool of messenger RNA (mRNA) from the samples (control and treated) is used to prepare fluorescently labeled complementary DNA (cDNA) by RT-PCR. To allow direct comparison between the control and treated samples, the samples are labeled with different fluorophores Cy3 (green) and Cy5 (red) respectively, for the control and treated samples. The labeled cDNA samples are then mixed and hybridized to a microarray with each spot on the array representing a distinct gene of interest. Comparison of relative signal intensity between the two labeled samples is then used to determine the relative differential expression for individual gene transcripts.

Alternatively, samples can be hybridized to separate but identical arrays. One of the often employed *in situ*-synthesized DNA microarray methods is the "GeneChips" technology by Affymetrix where the arrays are by photolithographic and combinatorial chemistry (Fodor *et al.*, 1991). In this case, each labeled cDNA population is then hybridized to separate but identical arrays. Inter-array comparisons are made using a global normalization factor that scales to average signal intensity on each array, and results can be presented in a false-color manner similar to the double color method. In cases where a large number of microarray comparisons are made, statistical analysis or data clustering of gene transcript families can then be performed.

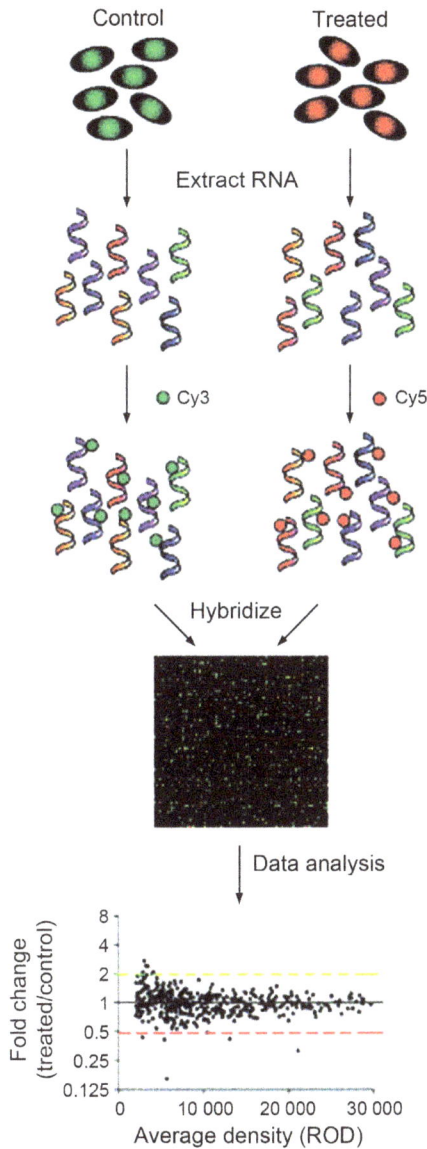

Figure 8.12.   Gene expression analysis using a DNA microarray. Complementary DNA (cDNA) extracted from the messenger RNA (mRNA) form two different samples (control and treated) and are fluorescently labeled with Cy3 and Cy4. When these target cDNA samples are hybridized on the probe DNA on chip (designed for a specific genome), the relative ratio of the green to red fluorescent signal can be directly used to measure the relative abundance of transcripts. Reproduced with permission from Marcotte *et al.* (2001).

DNA microarray technology has been extensively used in gene expression profiling (relative expression profiles of transcripts due to exposure to pathogens, the effect of treatment, the role of diseases), Chromatin immuno-precipitation on Chip (immunoprecipitating of protein along with a bound DNA sequence), and SNP detection (single nucleotide polymorphism).

8.2.5 *Nanoparticles*

Nanoparticles (NPs) can be of the "hard" or the "soft" varieties. The primary essence of NP is their small size, nanometer-scale particles. NPs of the hard variety include quantum dots (QD) whose size can determine the optical emission wavelength, and nano-gold particles (GNPs), whose emission can be enhanced via surface plasmonic resonance enhancement. Such field enhancement is useful for diagnostic procedures that utilize Raman scattering. Surface Enhanced Raman Scattering (SERS) signals are shown to have the capability of increasing the spontaneous Raman signal from $100\times$ to $10^{+14}\times$. Nanoparticles of the soft variety are mainly micelles and liposomes. In either of these configurations where lipid molecules are the extremal element, light-emitting elements such as rare earth particles or QDs can be encapsulated by one or two layers of lipids, using the process called self-assembled monolayers (SAM). Liposomes are particularly adaptable for delivering not only light-emitting entities but also therapeutic entities that can be captured in these sacs and released at the necessary sites for localized therapy, such as photodynamic therapy, which we shall cover in the next section. Coupling these diagnostic and therapeutic elements with endoscope allows the medical practitioner control in not only reaching regions of difficult access (joints and brain) but also provide means of diagnostics and therapy on those locations (Wilson and Borel, 2016). These authors also discuss the recent incorporation of nonlinear optical endoscopic methods. In that case, because near IR light penetrates tissue material more deeply, the use of IR excitation is highly advantageous. Furthermore, damage from photobleaching is also less pronounced using processes such as MPEF for diagnostic purposes.

8.2.6 *Fluorescence Lifetime Imaging Microscopy (FLIM)*

It is well-known that the lifetime of a given fluorophore is strongly dependent on the environment that the emitter resides. As such, using lifetime imaging can pinpoint domains of different environmental conditions.

Extending this concept to probing tissues, fluorescence lifetime imaging of molecules over spatial extent can help us to map out domains of that tissue that is undergoing changes. Using intrinsic fluorophores, the common tissue fluorophores are aromatic amino acids (Tyr, Trp, and Phe). Other enzyme metabolic co-factors such as flavin adenine dinucleotide (FAD) or lipid components like porphyrins also yield autofluorescence signals that will change in a lifetime if the environment is modified or structure degraded. Mapping these domains in real time provides a means of identifying diseased cells and tissues, allowing the physician to carry out the most appropriate therapeutic steps. Such a review was presented by Gorpas and Marcu (2016). The key advantage of this method is the ability to conduct *in situ* tissue diagnostics, searching for specific types of cancerous cells or otherwise damaged cells. Coupling these techniques with endoscopy further allows for tighter spatial resolution.

8.3 Photonics Means of Medical Therapy

8.3.1 *Photodynamic therapy (PDT)*

The idea of photodynamic therapy (PDT) is one of delivering a dose of optical radiation into a region within the body for the purpose of creating some localized cellular damage, preferably damage to the actively invading, disease-generating cells. The key requirement is to bring the delivery vehicle (molecule, vesicle, nanoparticle) to the cell or tissue region that has been deemed harmful to the health of the body with as little collateral damage as possible. Thus the element of specific cell recognition is very important. Once the vehicle has been positioned properly, the next important thing is to bring the optical radiation to the site of the vehicle and allow the photonic process to release the local poison. In order to minimize the collateral damage of the healthy cells and tissues surrounding this cluster of undesirable cells, concern has to be with how the poison takes effect. That is, does the poison simply poison anything in its neighborhood or is the released agent specific to the goal of eliminating the specific type of cells, e.g., cancerous cells? Let us review the science and technology of PDT.

Historically, PDT has been used most often on diseased cells within the vascular system, within the numerous body organs and within the macula of the eye. The type of diseased cell is usually related to cancer, and in the case of Age-related Macular Degeneration (AMD), diseased retinal cells. What this means is that in most cases, the location for the delivery is macroscopically discernable, and this location can be pinned down more tightly by the use of either fluorescence or Raman spectroscopy signatures, such that these signatures from the delivery agent can be measured spatially and visually. Endoscopic delivery and monitoring can be achieved for vascular lesions. From a review of the status of PDT by Wilson and Patterson (2008), it is interesting to note that the localization of the delivery agent has not been the primary focus of the field so far.

Delivery of light into the region of need is however, of great concern. As is well-known, the light wave is of the wavelength that will be strongly scattered and sometimes inadvertently absorbed by tissues even before the light reaches the location of PDT need. Thus much has been done to analyze the passage of light in dense, opaque medium, using what is called the Light Diffusion Equation. Here, random scattering of light allows for analysis of light propagation by the random diffusion process. Flows can alter the light propagation by creating Doppler shifts and broadening. To minimize the scattering of light, much research has been done to shift toward longer wavelength IR waves. These have longer penetration depth than the visible light sources. Lasers, LEDs, and even filtered arc lamps have been used. Usually between 1–5 W of usable power is required in the 630–850 nm range with an irradiance of a few hundred mWcm^{-2}. The key requirement is for the incident light source to have the desired light wavelength matching the photosensitizer for the release of the damaging species. A tunable light source has since become the norm. The array structure of tunable LEDs has become a popular user-configurable geometry to match the macroscopic treatment area.

Unlike gamma radiation therapy, the PDT process is designed to create damage of the cells in a manner that the cellular materials are still usable. That is, the process of cell death is not necrotic but apoptotic. Thus, in a manner of speaking, the dead cells are then recycled instead of simply becoming necrotic and excreted.

8.3.2 *Mechanism of PDT*

Figure 8.13 shows the schematic diagram of the basic photophysical processes in PDT. The sensitizer has the basic absorption band in the 560–800 nm range. Essentially this is the S_0 to S_1 transition. The sensitizer can fluoresce and return to the S_0 state or execute an intersystem crossing (ISC) to excite the associated triplet manifold in the T_1 state. An effective sensitizer will predominantly send the excited system into that triplet state.

This state has the spin state of 1 instead of 0, and thus cannot simply return to the ground S_0 state easily. However, in the presence of molecular oxygen $^3O_2$, which in its own ground state is already in the triplet state, this molecule can strongly interact with the T_1 state sensitizer molecule. In this reaction, the sensitizer returns to its ground state while producing singlet $^1O_2$ species. It is this reactive $^1O_2$ species that is primarily responsible for creating *oxidative damage* to the neighboring cells. The range of interaction is typically 20 nm surrounding the cytotoxic molecule, and its lifetime for the function is ~ 40 nanosecond. Figure 8.13 shows this approach of effecting prescribed cellular damage in this manner as Type II.

Shown also in Figure 8.13 is Type I pathway whereby *superoxide* is the cytotoxic species. In this pathway, superoxides O_{2-}^* are directly produced via collision with the O_2 molecule in the presence of free electron and the triplet state sensitizer. The fact that O_{2-}^* species can be as locally toxic when it converts to highly reactive hydroxyls radicals OH^* makes this process (Type I) an equally useful pathway for PDT therapy. This pathway is more commonly used for damaging fatty acids and lipids.

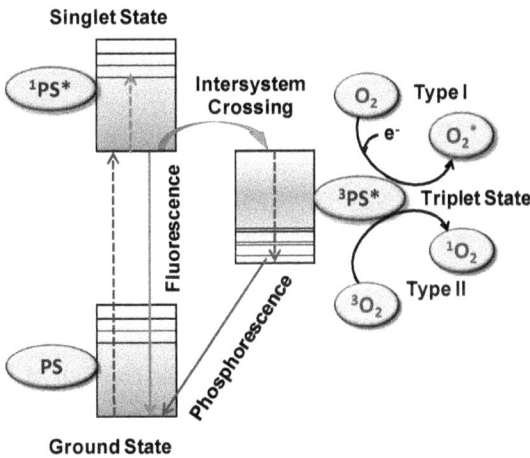

Figure 8.13. Photochemical mechanisms in PDT. Ground state photosensitizer molecule absorbs light that excites it to singlet state that can lose energy by fluorescence or can undergo intersystem crossing to long-lived polarization-sensitive triplet state that can carry out photochemistry or lose its energy by phosphorescence. Subsequently this photochemistry leads to local production of reactive oxygen species such as singlet oxygen (Type II) or superoxide (Type I) that are cytotoxic to microbial cells and to host cells. Figure remade using details from Wilson and Patterson (2008); Kharkwal *et al.* (2011); and Abrahamse and Hamblin (2016).

Table 8.1.   Clinical applications of PDT.

• Solid Tumors*	• Dysplasias*
○ Skin-BCC	• Age related macular degeneration*
○ Lung*	• Actinic Keratosis*
○ Esophagus*	• Papillomas
○ Bladder	• Rheumatoid Arthritis
○ Head & Neck	• Cosmesis
○ Brain	• Psoriasis
○ Ocular Melanoma	• Endometrial Ablation
○ Ovarian	• Localized Infections
○ Prostate	○ Fungal
○ Renal Cell	○ Bacterial
○ Cervix*	• Prophylaxis of Arterial Restenosis
○ Pancreas	
○ Bone	

*Approved photosensitizer/indication

Type of condition	Where found	PDT drug
Solid tumors	Lung, esophagus, bladder	Photofrin (HPD)
AMD	Eye	Visudyne
Actinic Keratosis	Skin	ALA

It has been shown that continuous wave (cw) laser light is more effective than the short-pulsed laser in delivering the PDT dosage. This turns out to be related to the fact that cells respond to the presence of $^1O_2$ species at its own molecular timescale. The short pulses simply will not create enough damaging oxidative species in the bursts of excitation.

PDT has been used extensively for solid tumors in various organs and vascular lesions. There seems to be a way to stop or at least slow down the progression of AMD. Due to low systemic toxicity, it has been widely used in treating skin lesions and systemic blood disorders. Table 8.1 shows a range of effective sensitizers for the treatment of specific diseases. A solid tumor in an upper airway can be cleared using Photofrin (HPD-PDT). Using Visudyne-PDT for AMD allows closure of neovasculature and the preservation of normal retinal vessels. The use of ALA-PDT treatment for skin actinic keratosis is also highly effective. The top half of Table 8.1 shows the vast range of use of PDT. Specifically, a list of solid tumors that have been treated is shown in blue, in descending order of the number of clinical trial patients reported.

8.3.3   *Emerging areas of research in PDT*

The current status of PDT has certain emerging areas of focus.

(1) Just like the use of cw lasers instead of short-pulse lasers, the future is also looking into slower, mini-dose PDT (mPDT) instead of a single,

large dose delivery. The idea is the same; create more species of the singlet O_2 ($^1O_2$) so that the oxidative species can be there for a longer duration, commensurate with the dynamics of the cells and tissues.

(2) The use of two-photon PDT for the excitation of the sensitizer is a way to create more localization, very much similar to the use of two-photon absorption to excite fluorophores for more localization visualization. Furthermore, the use of longer-wavelength photons allows for higher depth penetration in the tissue medium. So there is a need to compare the advantages of depth penetration and spatial localization achievable in two-photon PDT versus conventional cw PDT.

(3) A second type of two-photon scheme is to use a single photon to create the sensitizer excitation and ISC into the triplet state. Then a second photon excites the T_1 state into a T_n state, which in its own right can function as the photo-toxic species interacting with the biological molecule of intent, allowing more choice of created species to be the oxidative carrier (**Figure 8.14**). In this second scheme, it is shown that without molecular oxygen, the process can still work. Thus the use of the second laser creates a new pathway to produce oxidative species in anaerobic conditions.

(4) Molecular beacon — An idea of creating more directed dosage delivery is the use of molecular beacon to carry out a competitive process. When the sensitizer is in the S_1 state, if there were another acceptor fluorophore within its range, we recall that the FRET process is a way for energy to be transferred from the donor to the acceptor. In this

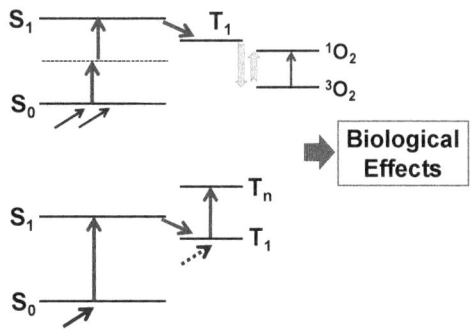

Figure 8.14. Two paths for two-photon excitation PDT. Toxic species is produced in the presence of $^3O_2$ (upper), or via T_n in oxygen-free environment created by second laser for T_1–T_n transition (lower). Adapted from Wilson and Patterson (2008).

case, the acceptor depletes the excited state of the sensitizer so that no ISC takes place, hence no sensitizer process is created. So if one adds another process to displace the acceptor molecule, the beacon acceptor light will disappear, and the excited S_1 state will undergo its normal ISC and lead to the oxidative species. This will lead to more controlled delivery of the damaging species.

(5) Nanoparticle delivery system — Nanoparticles are small entities such as a fullerene or other nanostructures including the lipoprotein structures. Using these small delivery vehicles can possibly create a more precisely directed region for the PDT to take place.

8.4 Controlling Molecular Gating Functions

In nature, light-activated controls of biological functions are plentiful. Through the action of phytochromes, plants seek to avoid shade that is depleted in the energy-rich red region of the spectrum. Plant growth response reflects the most obvious of the "shade avoidance" responses that include altered leaf architecture and early flowering (Duanmu *et al.*, 2013). Light can also be directed to activate functions of the human body such as twitching of a nerve cell, the trigger of a nerve cell, the trigger of a cardiac pulse or the release of an insulin molecule. The accurate control of these light-activated mechanism heralds a new field call *optogenetics*, with promise of molecular specificity in optically managed diseases such as Parkinson's Disease, cardiovascular accident, or Type II diabetes. The thrust is based on the fact that many of the human diseases have, in their molecular origin, the defective management of some triggering signal at the cellular level. The excessive rapidity of firing of some sets of nerve cells leads to Parkinson-type symptoms; the rate of involuntary cardiac muscle contraction such as atrial fibrillation has in its control the faulty Ca^{++} release in muscle cells; it is also known that the inadequate release of insulin in pancreatic cells may be blamed on G-protein receptor abnormalities. The idea of developing ways to control these necessary functions has its genesis in this new field of optogenetics.

The origin of this new field was based on an observation by Peter Hegemann in 1986, when he asked the question "Why do the algae swim toward the light?" (Hegemann and Moglich, 2011). His theory is that there may be a molecule within the single cell organism that senses light and manages the necessary sub-cellular dynamics to drive the cell toward more light. So that all-important element must be a light-sensing protein within the cell.

We then fast forward to 2003, when Hegemann and Georg Nagal successfully isolated a light-sensitive protein in algae. This is called *channelrhodopsin-2* (ChR2). The uniqueness of this protein is that it can be taken from the algae and transplanted into other cells, preserving much of its requisite functioning properties. The initial success in this transfection of the ChR2 into mammalian neuron cell was accomplished in Karl Deisseroth's laboratory in 2005 (Deisseroth, 2011). In November 2010, this ChR2 gene was also expressed in mouse heart muscle, allowing it to be controllably triggered by light. It is postulated that insulin-secreting beta cells can be put under similar control systems.

8.4.1 *Mechanism of action*

In the schematic cartoon in **Figure 8.15**, we consider the basic idea of *optogenetics*. If the light-sensitive receptor ChR2 has developed as the result of gene transfection, to be a specific, active transporting or gating protein

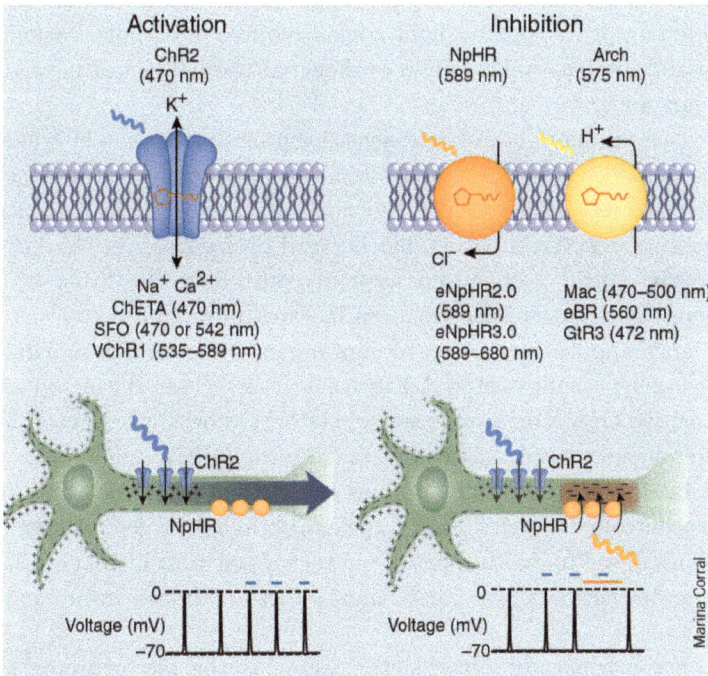

Figure 8.15. Optogenetic tools for modulating membrane voltage potential. Reproduced with permission from Pastrana (2011).

system on the membrane (left upper diagram), a channel protein system that actively controls the flux of K^+ and Na^+ ions now exists. Well-known is the fact that the imbalance of these ions will lead to the development of action potential for the nerve cell, causing it to fire. The introduction of specific, optically-controlled proteins, here ChR2 located in the light-sensitive domain of the ion channel or pump, now can be shown to have the ability to modulate neuronal signaling.

In the presence of blue light (450 nm), this non-selective cation channel immediately depolarizes the neurons and triggers a spike, as shown on the illustration in the lower left. Variants of this light-sensitive protein have been developed. Some of these are listed in the middle of the left side of this cartoon. For example, the step-function opsins (SFO) are slower versions of the ChR2, and can induce prolonged stable excitable states in neurons upon exposure by blue light, and can do the reverse upon excitation by green light (Figure 8.15, left).

On the right side of Figure 8.15, we show that the inhibition of neuronal triggering can be achieved by light stimulating of halorhodopsin (NaHR), a chloride pump. Here yellow light trigger induces hyperpolarization of the neurons and inhibits spiking. This is shown at the bottom of the right side of this cartoon.

The expanding toolbox of light-sensitive proteins that can be transfected with the genes of many channel proteins and G-protein coupled receptors (GPCR) has become the heart of this new approach to optically controlling the signaling of well-established biochemical pathways.

Recently, ChR2 was indeed used in controlling the firing of mouse heart cells. This is the initial success in extending optogenetics into non-neuroscience applications. A cartoon of one of these GPCR signaling processes using light is illustrated in **Figure 8.16(a)**. Here, the introduction of light-sensitive OptoXRs into adrenergic GPCR allowed the light at 505 nm to control signal cascades downstream from the GPCR.

Non-membrane associated proteins can also be manipulated to execute the light controlled activity. In Figure 8.16(b), light, oxygen, voltage (LOV) domains are fused to another protein in such a way that blue light (450 nm) can effect downstream activities. Among these activities are new binding pathways.

Another cytoplasmic optogenetic system is the phytochrome (PhyB) from the plants as shown in Figure 8.16(c), which can be triggered by red light to bind to its protein partner PIF (Phytochrome Interacting Factor).

This binding of PIF with PhyB allows fusion of PIF with the Rho-family GTPases, leading to localized actin recruitment and eventually, polymerization and formation of cellular extensions.

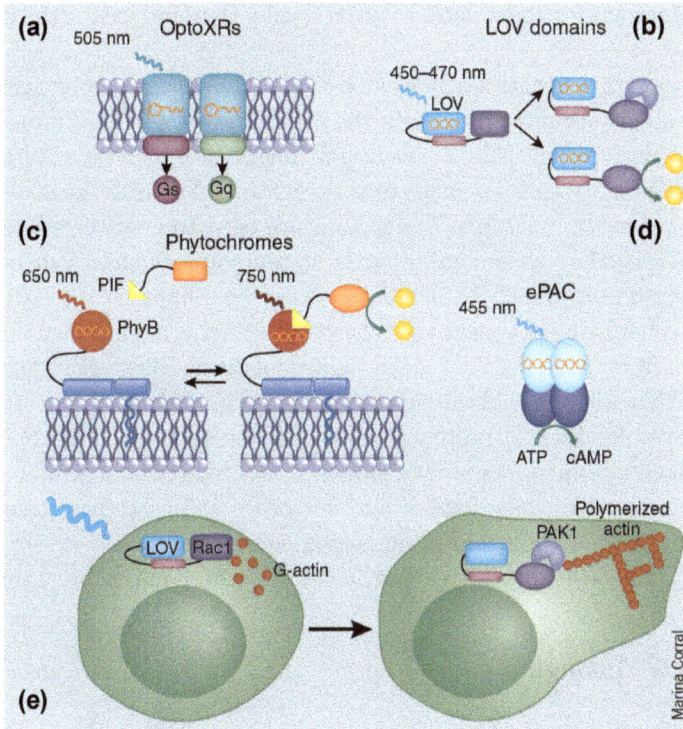

Figure 8.16.   Optogenetic tools for modulating intracellular signaling. (a) OptoXRs (Chimeric proteins) composed of bovine rhodopsin and the intracellular components of adrenergic G-protein coupled receptors allow optical control of G-protein mediated signaling cascades. (b) Non-membrane-associated photosensitive, light, oxygen, voltage (LOV) domains, phytochromes or cryptochromes from plants can be fused to cellular effector proteins to create light-sensitive variants. (c) An alternative cytoplasmic optogenetic system consists of the photoreceptor PhyB and its protein binding partner PIF. Red light triggers the binding of PIF to PhyB, and infrared light releases PIF. (d) Naturally occurring enzymes such as the light-activated adenylyl cyclase (ePAC), can be used to modulate cell signaling events by direct production of second messenger molecules. (e) Blue light induces a change in the conformation of the LOV2 domain, which results in the release of the allosteric block of Rac, allowing it to bind and activate downstream targets such as PAK1, leading to the polymerization of actin filaments, and the generation of localized cell protrusions and cell movement. Reproduced with permission from Pastrana (2011).

In yet another variant shown in Figure 8.16(e), the actin recruitment is brought about by similar actions but a different set of proteins. Here, the LOV2 domains are encoded with flavin as a chromophore. This activation by blue light changes the conformation of the LOV2 domain, resulting in a release of an allosteric block of Rac, which allows it to bind and activate downstream targets such as PAK1, and that induces actin filament formation.

To recap, optogenetics defines a class of techniques whereby genetically coded proteins are introduced into the cellular system. These proteins will effect the activity levels of downstream functions upon the absorption of light. The activity change comes in various forms. What we have described here are primarily microbial opsins that are used to control cellular functions. We see that in general, spatial or temporal changes can be made by the light absorption. In a recent report by Zhang *et al.* (2017), the list of mechanistic approaches for these optogenetic management has now extended from proteins that have light-dependent allosteric control and proteins that undergo light-dependent changes in oligomeric interaction to photocleavable (PhoCl) proteins. It is tempting to project the possibility that these temporal and spatial changes of selected proteins can be used to control downstream functions of another process. These then form a new repertoire of tools for possibly managing some of the most severe human diseases in the near future. We look forward to those major breakthroughs.

8.5   Final Thoughts

The application of novel photonic tools to help in the search for cures and prevention in medical practices is moving rapidly. The continuing aspiration of all medical practitioners is a single purpose: to detect diseases at their earliest stage and make adjustments to prevent or retard the development of diseases. Recent innovations associated with the Clustered Regularly Interspaced Short Palindromic Repeats (CRISPR) research efforts in genome editing for the elimination of diseased genes in human embryos are earth-shaking and groundbreaking. It is conceivable that both optical monitoring with ultrahigh resolution microscopy as well as optogenetics approaches can be major players in the solution of our important medical problems using the CRISPR approach. The physical and biological scientists, as well as engineers of all different persuasions, continue to explore newer and more efficient avenues to arrive at the goals set by the medical field. Biophotonics will continue to be at the forefront of the field of biomedicine, pushing the envelope of research and development for newer and faster approaches for diagnosis and prevention of diseases. These lectures provide only the very rudimentary basics of the ever-expanding field. This is a field ripe for new innovations and creativity.

References

Abrahamse, H. and M.R. Hamblin. New photosensitizers for photodynamic therapy. *Biochem. J.* 473: 347–364, 2016.

Balhorn, R., S. Hok, P.A. Burke, F.C. Lightstone, M. Cosman, A. Zemla, G. Mirick, J. Perkins, A. Natarajan, M. Corzett, S.J. Denardo, H. Albrecht, J.P. Gregg and G.L. Denardo. Selective high-affinity ligand antibody mimics for cancer diagnosis and therapy: Initial application to lymphoma/leukemia. *Clin. Cancer Res.* 13: 5621s–5628s, 2007.

Brown, P.O. and D. Botstein. Exploring the new world of the genome with DNA microarrays. *Nat. Genet.* 21: 33–37, 1999.

Chen, Z. and S. Tang. Second harmonic OCT and Combined MPM/OCT. In: W. Drexler and J.G. Fujimoto (eds.) *Optical Coherence Tomography*, Springer International Publishing, Switzerland, 2015, pp. 1489–1514.

Chen, Z., Y. Zhao, S.M. Srinivas, J.S. Nelson, N. Prakash and R.D. Fostig. Optical Doppler Tomography. *IEEE J. Sel. Top. Quantum Electron.* 5: 1134–1142, 1999.

De Montigny, E., W.J. Madore, O. Ouellette, G. Bernard, M. Leduc, M. Strupler, C. Boudoux, C. and N. Godbout. Double-clad fiber coupler for partially coherent detection. *Opt. Express.* 23: 9040–9051, 2015.

Deisseroth, K. Optogenetics. *Nat. Methods* 8: 26–29, 2011.

Delano, W.L. PyMOL. The PyMOL Molecular Graphics System, Version 1.2r3pre, Schrödinger LLC, 2002.

Diehnelt, C.W., M. Shah, N. Gupta, P.E. Belcher, M.P., Greving, P. Stafford and S.A. Johnston. Discovery of high-affinity protein binding ligands — backwards. *PLoS One* 5: e10728, 2010.

Fercher, A.F. Optical coherence tomography — development, principles, applications. *Zeitschrift für Medizinische Physik* 20: 251–276, 2010.

Fodor, S., J. Read, M. Pirrung, L. Stryer, A. Lu and D. Solas. Light-directed, spatially addressable parallel chemical synthesis. *Science* 251: 767–773, 1991.

Gorpas, D. and L. Marcu. Fluorescence lifetime spectroscopy and imaging techniques in medical applications. In: M. Olivo and U.S. Dinish (eds.) *Frontiers in Biophotonics for Translational Medicine*, Springer+Business Media, Singapore, 2016, pp. 1–46.

Hegemann, P. and A. Moglich. Channelrhodopsin engineering and exploration of new optogenetic tools. *Nat. Methods* 8: 39–42, 2011.

Ji, N. Adaptive optical fluorescence microscopy. *Nat. Methods* 14: 374–380, 2017.

Jonnal, R.S., O.P. Kocaoglu, R.J. Zawadzki, Z. Liu, D.T. Miller and J. Werner. A review of adaptive optics optical coherence tomography: Technical advances, scientific applications, and the tuture. *Invest. Ophthalmol. Vis. Sci.* 57: OCT51–68, 2016.

Kharkwal, G.B., S.K. Sharma, Y.Y. Huang, T. Dai and M.R. Hamblin. Photo-dynamic therapy for infections: Clinical applications. *Lasers Surg. Med.* 43: 755–767, 2011.

Krishnan, V.V., I.H. Khan and P.A. Luciw. Multiplexed microbead immunoassays by flow cytometry for molecular profiling: Basic concepts and proteomics applications. *Crit. Rev. Biotechnol.* 29: 29–43, 2009.

Krishnan, V.V., I.H. Khan and P.A. Luciw. Biomarker detection and molec-ular profiling by multiplex microbead suspension array based immuno-proteomics. In: R. Rapley and S. Harbron (eds.) *Molecular Analysis and Genome Discovery*, 2nd edn., John Wiley & Sons, New York, 2012, pp. 244–270.

Liu, R., X. Li and K.S. Lam. Combinatorial chemistry in drug discovery. *Curr. Opin. Chem. Biol.* 38: 117–126, 2017.

Marcotte, E.R., L.K. Srivastava and R. Quirion. DNA microarrays in neuropsy-chopharmacology. *Trends Pharmacol. Sci.* 22: 426–436, 2001.

Mo, G.C., B. Ross, F. Hertel, P. Manna, X. Yang, E. Greenwald, C. Booth, A.M. Plummer, B. Tenner, Z. Chen, Y. Wang, E.J. Kennedy, P.A. Cole, K.G. Fleming, A. Palmer, R. Jimenez, J. Xiao, P. Dedecker and J. Zhang. Genetically encoded biosensors for visualizing live-cell biochemical activity at super-resolution. *Nat. Methods* 14: 427–434, 2017.

Pastrana, E. Optogenetics: Controlling cell function with light. *Nat. Methods* 8: 24–25, 2011.

Qu, Y., T. Ma, Y. He, M. Yu, J. Zhu, Y. Miao, C. Dai, P. Patel, K.K. Shung, Q. Zhou and Z. Chen. Miniature probe for mapping mechanical properties of vascular lesions using acoustic radiation force optical coherence elastogra-phy. *Sci. Rep.* 7: 4731, 2017.

Reslova, N., V. Michna, M. Kasny, P. Mikel and P. Kralik. xMAP technology: Applications in detection of pathogens. *Front. Microbiol.* 8, 2017.

Su, J., I.V. Tomov, Y. Jiang and Z. Chen. High-resolution frequency-domain second-harmonic optical coherence tomography. *Appl. Opt.* 46: 1770–1775, 2007.

Van Velthoven, M.E.J., D.J. Faber, F.D. Verbraak, T.G. Van Leeuwen and M.D. De Smet. Recent developments in optical coherence tomography for imaging the retina. *Prog. Retin. Eye Res.* 26: 57–77, 2007.

Wilson, B.C. and S. Borel. Nanoparticle-enabled optical endoscopy: Extend-ing the frontiers of diagnosis and treatment. In: M. Olivo and U.S. Din-ish (eds.) *Frontiers in Biophotonics for Translational Medicine*, Springer Science+Business Media, Singapore, 2016, pp. R16–R109.

Wilson, B.C. and M.S. Patterson. The physics, biophysics and technology of pho-todynamic therapy. *Phys. Med. Biol.* 53: R61–109, 2008.

Zhang, W., A.W. Lohman, Y. Zhuravlova, X. Lu, M.D. Wiens, H. Hoi, S. Yaganoglu, M.A. Mohr, E.N. Kitova, J.S. Klassen, P. Pantazis, R.J. Thompson and R.E. Campbell. Optogenetic control with a photocleavable protein, PhoCl. *Nat. Methods* 14: 391–394, 2017.

Index

ablation, 254
acceptor, 41, 63, 74, 197, 225, 255
acetylcholine, 176, 190, 226
A-coefficient, 38
acoustic, 231, 262
  acoustical, 235, 236
  AO (acousto-optical), 240
actin, 89, 90, 110–114, 127, 157, 198–200, 259, 260
adaptive optics
  adaptive optical corrections, 240
  adaptive optical fluorescence, 261
  adaptive optical lenses, 240
  adaptive optical methods, 240
  adaptive optics, 240, 262
  adaptive optics microscopy, 240
adenylyl cyclase, 259
  Adenylyl-imidodiphosphate, 112
adipocyte, 180.
ADP (adenosine diphosphate), 112, 114, 128
Aequorea Victoria, 4, 146
aequorin, 146
Affymetrix, 248
AFGP (Antifreeze Glycoprotein), 92
AFM (Atomic Force Microscopy), 101, 102, 124, 140
aggregate, 78
  aggregation processes, 200
agonism, 109
Airy, 130–132, 137, 150, 182, 183, 202

ALA-PDT (see Photodynamic therapy (PDT)), 254
Alexa, 144, 207, 216, 247
  Alexa-488-telenzepine, 145
  Alexa-647, 220
algae, 5, 256
allosteric, 81, 89, 259, 260
ALS (Advanced Light Source), 183
Alzheimer's, 8, 9
AMD (Age-related macular degeneration), 251, 254
AMP-PNP (Adenylyl-imidodiphosphate), 112
anemometry, 201
angular
  angular distance, 131
  angular intensity, 33
  angular magnification, 129
  angular orientation, 41
  angular radius, 130, 131
  angular rotation, 121
  angular scan, 30, 34
  angular separation, 131
  angular spread, 131
anisotropic, 111, 168
  anisotropy, 23, 25, 40, 126, 167
annular, 214
  annulus, 134
anomalous, 21, 70, 182, 194, 195
anti-bunching, 217–221, 227
anti-Stokes, 36, 45–47, 174